Paul M. S. Monk, Roger J. Mortimer,
David R. Rosseinsky

Electrochromism:
Fundamentals and Applications

VCH

Further Titles of Interest by VCH

H. Gerischer, C. W. Tobias (Eds.)
Advances in Electrochemical Science and Engineering

Volume 1 ISBN 3-527-27884-2
Volume 2 ISBN 3-527-28273-4
Volume 3 ISBN 3-527-29002-8
Volume 4 ISBN 3-527-29205-5

J. Lipkowski, Ph. N. Ross (Eds.)
Frontiers of Electrochemistry
Volume 3. Electrochemistry of Novel Materials

ISBN 0-89573-788-4

J. Wang
Analytical Electrochemistry

ISBN 1-56081-575-2

Distribution:
VCH, P. O. Box 10 11 61, D-69451 Weinheim (Federal Republic of Germany)
Switzerland: VCH, P. O. Box, CH-4020 Basel (Switzerland)
United Kingdom and Ireland: VCH (UK) Ltd., 8 Wellington Court, Cambridge CB1 1HZ (England)
USA and Canada: VCH, 220 East 23rd Street, New York, NY 10010-4606 (USA)
Japan: VCH, Eikow Building, 10–9 Hongo 1-chome, Bunkyo-ku, Tokyo 113 (Japan)

ISBN 3-527-29063-X

Paul M. S. Monk, Roger J. Mortimer,
David R. Rosseinsky

Electrochromism: Fundamentals and Applications

VCH Weinheim • New York • Basel • Cambridge • Tokyo

Dr. P. M.S. Monk
Department of Chemistry
Manchester Metropolitan University
Chester St.
Manchester M1 5GD
UK

Dr. R. J. Mortimer
Department of Chemistry
Loughborough University of Technology
Loughborough
Leicestershire LE11 3TU
UK

Dr. D. R. Rosseinsky
Department of Chemistry
University of Exeter
Stocker Road, Exeter
Devon, EX4 4QD
UK

Published jointly by
VCH Verlagsgesellschaft mbH, Weinheim (Federal Republic of Germany)
VCH Publishers, Inc., New York, NY (USA)

Editorial Directors: Dr. Peter Gregory, Dr. Ute Anton
Production Manager: Dipl.-Ing. (FH) Hans Jörg Maier

Cover illustration: The image of a building is formed by electrochromic heptylviologen on a 1 inch (2.54 cm) square, 64 x 64 pixel, silicon electrode. (D. J. Barclay and D. H. Martin in E. R. Howells (ed.), *Technology of Chemicals and Materials for the Electronics Industry*, Ellis Horwood, Chichester **1984**, Chapter 15. Used with the kind permission of E. R. Howells. The electrochromic rearview mirror for a car was kindly supplied by Dr. H. Byker, Gentex Corporation, Zeeland, MI, USA.

Library of Congress Card No. applied for.

A catalogue record for this book is available from the British Library.

Deutsche Bibliothek Cataloguing-in-Publication Data:
Monk, Paul M. S.:
Elektrochromism : fundamentals and applications / Paul M. S. Monk ; Roger J. Mortimer ; David R. Rosseinsky. - Weinheim ; New York ; Basel ; Cambridge ; Tokyo : VCH, 1995
ISBN 3-527-29063-X
NE: Mortimer, Roger J.:; Rosseinsky, David R.:

Printed on acid-free and chlorine-free paper.

Printing: betz-druck gmbh, D-64291 Darmstadt. Cover design: Graphik & Text Studio Zettlmeier-Kammerer, D-93164 Laaber-Waldetzenberg.
Printed in the Federal Republic of Germany.

Preface

The field of electrochromism has changed much since the idea of an electrochromic display was first suggested in 1969. The introduction of liquid-crystal displays has necessitated a sharp change of focus. The applications originally envisaged for electrochromic devices (ECDs) usually relied on a rapid response in for example high definition television or optical computers. Applications at present considered suitable for ECDs are large-area displays, such as notice boards for traffic or for transport termini, the electrochrome being utilised against a reflective background; other light modulators act in a transmissive sense and will comprise, for example, a thin electrochromic device covering one side of a whole window. This latter aim, the construction of the so-called 'smart window', is a major technological goal.

There have been many previous reviews of electrochromism. Such works either tackle the topic from a more-or-less applied angle, for example covering one type of application, or concentrating on a single electrochrome. There has not hitherto been a monograph dedicated solely to the whole subject of electrochromism. The present work, while not intended to include all citations – there are many thousand – is the first to give a complete overview of the whole subject. Because of the size of the literature, in compiling this monograph almost arbitrary selections were required, and a cut-off date of late summer 1994 became inevitable. In our view, any treatment of electrochromism must include the underlying science, some of which might, at first sight, be considered rather special: however, such basic treatments have generally proven invaluable in the understanding of electrochromic phenomena. We have also, where suitable, included 'hands on' detail not found elsewhere, which may be useful to those entering the field.

Most of the science underlying electrochromism here is presented from a chemical viewpoint since electrochromism is an electrochemically-induced colour change. We have, however, endeavoured to make the exposition accessible to physicists or materials scientists and engineers. Thus, most chapters contain a few references imparting general background information if needed, but we have nevertheless probably erred by assuming either too little or too much prior knowledge.

This work is divided into three sections. Part I provides a general background for readers perhaps unfamiliar with the field. We include elementary definitions such as that for colouration efficiency, which are well known to the electrochromism community but for which an actual definition is rather hard to come by. Some basic electrochemical theory is included also. Part I concludes with a section on the construction of ECDs.

Part II describes both inorganic and organic chemical systems being considered at present for use in electrochromic applications. Chemical systems are presented approximately alphabetically.

Part III presents recent elaborations of electrochromism in some present-day research. The elaborations comprise polyelectrochromism and photoelectrochromism (including a discussion of electrochromic printing).

The production of a work such as this relies on the help and goodwill of many, and we wish to acknowledge the help and support of the following. First, we thank Dr Ute Anton of VCH for her editorial expertise and advice. We thank Manju Merjara of the Chemistry Department, MMU, for typing some of the original manuscript.

Besides providing extensive computer know-how and type-setting expertise, Joe Russell of the MMU helped reproduce many of the figures.

Figures have been reproduced by kind permission of the copyright holders, as follows: Butterworths (Fig. 4.4), Chapman and Hall (Fig. 12.1), The Electrochemical Society (Fig. 4.1), Dr E.R. Howells (Fig. 8.5), Elsevier (4.3 and 8.3), The Royal Society of Chemistry (Figs. 6.2, 8.2, 12.4 and 12.5) and the Society of Applied Spectroscopy (for Fig. 8.4).

We have had many helpful and stimulating discussions with other workers in the electrochromism community, in particular with Dr John Duffy, Dr Richard Hann, Professor Malcolm Ingram, the late Dr J. Brian Jackson, and Dr Robert Janes, Dr Poopathy Kathirgamanathan and Dr Andrew Soutar.

While the above have helped in producing this book, any errors remaining are ours.

P.M.S.M.	R.J.M.	D.R.R.
Manchester	*Loughborough*	*Exeter*

1995

Paul M. S. Monk is a lecturer in Physical Chemistry at the Manchester Metropolitan University. In 1990, he received his Ph.D. in chemistry from the University of Exeter having studied the electrochemistry of bipyridilium redox species. He then held a post-doctoral fellowship at the University of Aberdeen (1989–1991) performing research on rapid-response electrochromic devices based on tungsten-trioxide. His present research interests are mixed-metal oxide thin films for electrochromic purposes, novel (chiral) polyanilines and the effects of charge-transfer complexation on electron-transfer rates.

Roger J. Mortimer is a lecturer in Physical Chemistry at Loughborough University of Technology. In 1980, he received his Ph.D. from Imperial College having studied heterogeneous catalysis at the solid-liquid interface. He then held a post-doctoral fellowship at the California Institute of Technology performing research on polymer-modified electrodes. After a demonstratorship at the University of Exeter and lecturing positions in Cambridge and Sheffield, he took up his present post in 1989. His present research interests include electrochromism, electrochemical and optical sensors, and electrocatalysts for fuel cells.

David R. Rosseinsky has been reader in Physical Chemistry at the University of Exeter for as long as he can remember ('in the midst of life, we are in Exeter'). After M.Sc. research (Rhodes, South Africa) on electrolyte conductivities, a Ph.D. (Manchester) on aquo ion electron transfer, and a postdoctorate (University of Pennsylvania) effecting unintended siloxane-based explosions, two year's lecturing slog at the University of the Witwatersrand, South Africa, followed. Eyed up by Exeter during a further 3 year's postdoc (I.C.I. and Leverhulme), he was ultimately deemed fit for human consumption and appointed lecturer. He employs electrochemical probes in a wide variety of charge transfer processes: electron transfer rates in mixed valent solids, electrochemical photovoltaism, electrochromism, colloid electrodeposition, electropolymerisation, zinc-oxide electrophotography, composite electrostatic-charge acquisition, and high-temperature superconductors probed by liquid-phase electrochemistry around 100 K.

Contents

List of Tables

Symbols and Abbreviations

Symbols

A	area of electrode
A	absorbance ('optical density')
A	ampere
C	Coulomb
c	concentration of dissolved species
CR	contrast ratio
D	diffusion coefficient; chemical diffusion coefficient
E	potential of electrode (either impressed potential or zero-current potential)
E^{o}	standard electrode potential
E_{OC}	open-circuit (zero current) potential
$E_{1/2}$	half-wave potential
$E_{pa(m)}$	potential of *m*th anodic peak in CV
$E_{pc(m)}$	potential of *m*th cathodic peak in CV
F	Faraday constant
h	Planck constant
i	current
I	intensity of transmitted light
j	flux
J	Joule
k	rate constant
k_B	Boltzmann constant
K	equilibrium constant
K_{sp}	solubility product
l	thickness
L	the Avogadro constant
n	number of electrons involved in electron-transfer reaction
n	as subscript – a number of groups or atoms in a formula
Q	charge per unit area
R	gas constant
s	second
T	thermodynamic temperature
T	transmittance
μ	ionic mobility

υ	velocity of ion
V	volt
x	(as subscript) a number, often fractional, of atoms (ions) in a formula
x	(on par) a number, often fractional, of atoms (ions) in a reaction
x	insertion coefficient (consistent with the above)
$y,\ z$	(used as x, preceding)
z	charge number on ion
(M	abbreviation for the units mol dm^{-3})
α	electrochemical transfer coefficient (symmetry factor)
{ α	linear absorption coefficient (for optical absorption by solid species)
{ ε	extinction coefficient (molar absorptivity for species in solution)
ϕ	F/RT
{ η	colouration efficiency
{ η_{O}	overall colouration efficiency of electrochromic device
{ η_p	colouration efficiency of primary electrochrome
{ η_s	colouration efficiency of secondary electrochrome
ν	scan rate in cyclic voltammetry
ν	frequency of light
τ	response time; timescale
λ	wavelength
Ω	Ohm

Apparatus, Processes and Techniques

Abs	absorbance (optical density)
AC	alternating current
CE	counter electrode
CRT	cathode ray tube
CV	cyclic voltammogram
CVD	chemical vapour deposition
CT	charge transfer
DC	direct current
EBS	electron beam sputtering
EC	electrochromic
ECD	electrochromic device
EDAX	energy dispersive analysis of X-rays

ESR	electron-spin resonance
ET	electron transfer
FTIR	Fourier-transform infra red
IR	infra red
LCD	liquid-crystal display
LED	light-emitting diode
OTE	optically-transparent electrode
QCM	quartz-crystal microbalance
RE	reference electrode
RF	radio frequency
SCE	saturated calomel electrode
SEM	scanning electron microscope or micrograph
SHE	standard hydrogen electrode
SIMS	secondary-ion mass spectrometry
UV	ultra violet
WE	working electrode
XPS	X-ray photoelectron spectroscopy
XRD	x-ray diffraction

Materials

{ AIROF	anodically formed iridium oxide film
{ SIROF	sputtered iridium oxide film
AMPS	2-acrylamido-2-methylpropanesulphonic acid (polyAMPS is the derived polymer)
(aq)	aquo ion
{ $bipm^{2+}$	bipyridilium dication
{ $bipm^{+}$	bipyridilium radical cation
{ $bipm^{o}$	neutral bipyridilium-derived species
CPQ	cyanophenyl paraquat (1,1'-*bis*(*p*-cyanophenyl)-4,4'-bipyridilium)
DMF	dimethylformamide
e^{-}	electron
EtOH	ethanol
(g)	gaseous state [*cf.* (*l*) and (s)]
GC	glassy carbon
HCF	hexacyanoferrate

HV	heptyl viologen (1,1'-*n*-diheptyl-4,4'-bipyridilium)
MV	methyl viologen (1,1'-dimethyl-4,4'-bipyridilium)
ITO	indium tin oxide
(*l*)	liquid state [*cf.* (g) and (s)]
L	ligand
M	metal electrode; general metal or cation M^+ or M^{z+}
Me	methyl
MeCN	acetonitrile
MeOH	methanol
MV	methyl viologen (1,1'-dimethyl-4,4'-bipyridilium)
n	electron as negative charge carrier in solid
naph	naphthalocyanine
OP	osmium purple (iron(III) hexacyano-osmate(II))
p	positive hole as charge carrier through solid
PB	Prussian blue
s-PB	'soluble' Prussian blue
i-PB	'insoluble' Prussian blue
PG	Prussian green
PW	Prussian white
PX	Prussian brown (yellow in thin-film form)
pc	phthalocyanine
PC	propylene carbonate
PEO	poly(ethylene oxide)
Ph	phenyl
Pr	propyl
PVP	poly(vinyl pyrrolidone)
Q	quinone moiety
R	substituent
RP	ruthenium purple (iron(III) hexacyanoruthenate(II))
(s)	solid state [*cf.* (g) and (*l*)]
SIROF	sputtered iridium oxide film
TA	thiazine
TEAP	tetra-*n*-ethylammoniumperchlorate
TEAT	tetra-*n*-ethylammonium tetrafluoroborate
TPAP	tetra-*n*-butylammonium perchlorate

TBAT	tetra-*n*-butylammonium tetrafluoroborate
TCNQ	tetracyanoquinodimethane
TTF	tetrathiafulvalene
TSpc	tetrasulphonated phthalocyanine
X	general anion

Part I

Introduction

1 Electrochromism: Terminology, Scope, Colouration

1.1 What is Electrochromism?

An electroactive species often exhibits new optical absorption bands (i.e. shows a new colour) in accompaniment with an electron-transfer or 'redox' reaction in which it either gains or loses an electron; that is to say, it undergoes *reduction* or *oxidation*. Such colouration was first termed 'electrochromism' in 1961 by Platt [1] whose discussions were amongst the first published. Byker has discussed the historical development of electrochromism [2].

Many simple species exhibit electrochromism. To take a laboratory example, the ferrocyanide ion in aqueous solution is pale yellow in colour, but on electrochemical oxidation (loss of electron to an electrode):

$$\underset{\text{(pale)}}{[Fe^{II}(CN)_6]^{4-}} \rightarrow \underset{\text{(yellow)}}{[Fe^{III}(CN)_6]^{3-}} + e^{-}_{\text{electrode}}$$

a pool of brilliant yellow forms around the electrode, and thence diffuses into the bulk. The change in colour is directly attributable to the oxidation of iron(II) to iron(III) in the complex. A somewhat different case is ferrous ion in aqueous solution, in the presence of thiocyanate with which Fe^{2+} is only weakly complexed. Initially the solution is colourless, but a brilliant blood-red colour appears after oxidation on the formation of electro-generated iron(III). In this case, the colour may not be directly electro-generated, but is possibly due to interaction between electro-generated Fe^{3+} and the electro-inactive CNS^- ion in solution: it is the iron(III) thiocyanate charge-transfer complex that ultimately provides the colour. In this context, a 'charge-transfer' species is one in which a photo-effected transfer of charge within the species, sometimes between species, evokes colour, by 'optical charge transfer'.

A quite different system comprises an iron bathophenanthroline complex [3] *tethered* to a polymeric fluorocarbon support on an electrode in which reduction (electron gain) generates an intensely coloured iron(II) species.

Organic systems such as bipyridiliums (**I**) (also known as viologens or paraquats) can become highly coloured on reduction, again owing to intense optically-effected intramolecular charge transfer in the product. Species (**I**) have been studied, for n = 0–12.

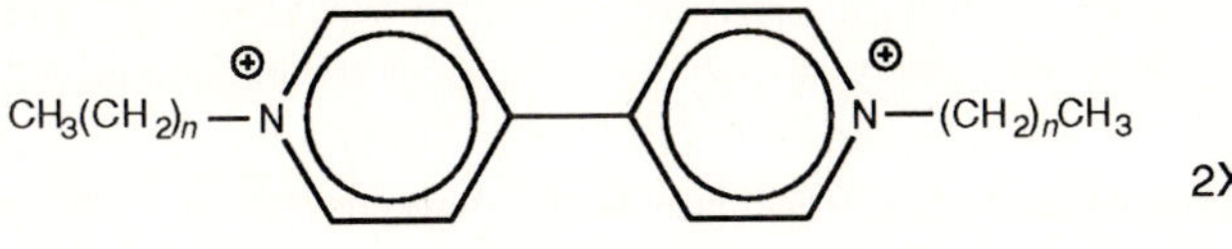

I

The most widely studied inorganic system is solid tungsten trioxide WO_3, also called tungsten oxide or tungstic oxide, comprising W^{VI}, in which the introduction of small amounts of W^{V} to give M_xWO_3 (M is a cation) again allows intense optical absorption or, with particular values of x in this case, reflection.

Generally, apart from the *conductive electrode* (metal or conducting glass), electrochromic species can be all liquid (e.g. the ferro-ferricyanide system cited), or all solid, as a film (the tethered iron bathophenanthroline system, or WO_3), or it can undergo liquid-to-solid conversion following oxidation or reduction (as in some bipyridilium systems). Where the electrochromic film is solid, oxidation is necessarily accompanied by anion incorporation from surrounding electrolyte, or cation expulsion from the film, while reduction will involve cation incorporation or anion expulsion. The transferring ion is called the *counter ion*. The ionic *diffusion* involved here, or the intrinsic *rate of electron uptake or loss*, will determine the rate of electrochromic operation. The general electrochemical operation of these systems is outlined in chapter 2.

The species that becomes coloured during redox reaction is sometimes called the *electrochromophore* or *electrochrome* [4] (see section 1.4.2). After the pulse of current effecting electron transfer at the working electrode has evoked the colouration, the colour persists, thereby producing the memory effect referred to below. Current in the opposite direction reverses the electrochemical process and the display reverts to the colourless or bleached state.

In our treatment, the term 'electrochromism' does not include phenomena such as shifts in optical band maxima induced by the application of high voltages (the Stark effect) on effectively immobilised molecules, for example, phenol blue in polystyrene [5]. Similarly, the electrochemical evocation of colour centres (F-centres and their myriad sub-species) in alkali and alkaline-earth halides [6] is excluded from consideration.

1.2 Existing Technologies

Display technology currently comprises cathode ray tubes (CRTs), liquid crystal displays (LCDs) and light emitting diodes (LEDs), and now obsolete discharge-tube elements.

The cathode ray tube can produce images of great clarity and complexity in many colours, as in television. Images can change rapidly and appear to move smoothly, resulting from the fast response. However, the CRT suffers disadvantages. Thus the CRT must withstand a high vacuum. Furthermore, high-energy electron sources have a large power consumption. The electron gun behind the screen becomes progressively longer as the screen become larger, and wide-angle viewing is difficult because of screen curvature. The screen pigments (phosphors) are expensive rare-earth compounds, and the necessary precision of device assembly make CRTs costly. Despite all this, relatively low-cost manufacture has been achieved by the sheer scale of production.

A second type of display depends on liquid-crystal technology. LCDs are flat and consume little power compared with the CRT, and the cost of system manufacture is also much lower. The LCD image is sharp with excellent clarity, although external lighting is usually necessary since the displays are 'passive', that is, not light emitting. An important requirement of LCD technology is the need for the glass display front face to be exactly parallel with the back plate for uniformity of field, a relatively easy requirement to meet for a small display, but more difficult as the display becomes larger [7]. Large-area LCD displays also involve difficulties in addressing a large number of picture elements, or 'pixels' [7]. Of necessity, LCD displays produce monochrome images, so stippling with dots is the only method available for tonal gradation if block colour is undesirable. Image persistence requires a constant power input since LCD displays have no inherent memory. Colour LCD devices are still comparatively rare and expensive.

LEDs are devices which include p,n junctions. In outline, semiconductors have bonding electrons in energy levels comprising the *valence band*, while at higher energies, suitable orbitals form a vacant *conduction band*, both bands pervading the space of the solid without overlapping. A p-doped semiconductor contains acceptor species with values of condensed-phase electron affinity so as to (just) abstract electrons from the valence band in which the remnant positive hole is then the charge carrier; n-doped semiconductors have donor species which (just) ionise electrons into the higher energy conduction band. A junction of such regions has unidirectional, rectifying, effects on the passage of current. With suitable populations of dopants, driving electrons by appropriate applied potentials across such junctions results in the recombination of electrons and holes which is accompanied by quite intense light emission. The red number-indicator glow on many instrument panels is of this kind. Electroluminescence is a comparable phenomenon, in which electrons forced into phosphors such as modified zinc sulphide cause impact ionisation and excitation of impurities, resulting in photon emission. Recently, novel LEDs comprising conductive organic polymers have been described [8]; hitherto, LEDs have been solely inorganic.

1.3 Electrochromic Displays and Shutters

Besides displays, electrochromic systems find an entirely novel application as optical shutters. Although electrochromic systems as displays need to compete with both CRT and LCD displays for commercial viability, they possess many advantages over both. Firstly, electrochromic devices (ECDs) consume little power in producing images which, once formed, persist with little or no additional input of power, in the so-called 'memory effect'. Secondly, there is no limit in principle to the size an ECD can take: a larger electrode expanse or a greater number of small electrodes [9] may be used. Multiple electrodes (pixels) allow text or images to be displayed rather than blocks of colour.

Tonal variation may be achieved by stippling with dots as with LCD displays, but the image may also be intensified by passing more charge into specified areas where more coloured substance hence is formed. There is however the technical problem with large area ECDs that patchy areas form when the current distribution is uneven across the electrode surface. An ECD may be either flat or curved for wide-angle viewing.

The ECD can be polyelectrochromic if the active component responds to different potentials with a variety of colours. Alternatively, pixels containing different electro-active species may be used. Recently Yasuda *et al.* [10] produced a trichromic ECD in which the red colour was formed from 2,4,5,7-tetranitro-9-fluorenone; green colour was formed as a product from 2,4,7-trinitro-9-fluorenylidene malononitrile and a blue colour was formed by electron transfer to TCNQ (tetracyanoquinodimethane). A device using Prussian blue and methyl viologen (1,1'-dimethyl-4,4'-bipyridilium) has been shown to evince five discrete electrochromic colours [11] and seven colours in a system comprising an electrode surface modified with polymeric *tris*(5,5'-dicarboxyester-2,2'-bipyridine) ruthenium(II) [12]. Polyelectrochromism is treated separately in chapter 11.

There are many disadvantages associated with EC displays: external lighting is needed for image visibility under certain conditions and, since many ECDs contain liquid electrolytes, there are possible problems of construction and storage (see chap. 3). At present there are operational difficulties with most ECD prototypes, although several devices are now available commercially, for example, a WO_3 based device has been used as a display to indicate the price of shares in the Tokyo stock exchange [13], and a liquid-phase bipyridilium/thiazine or phenylenediamine system is employed in an automatically darkening rear-view mirror [14].

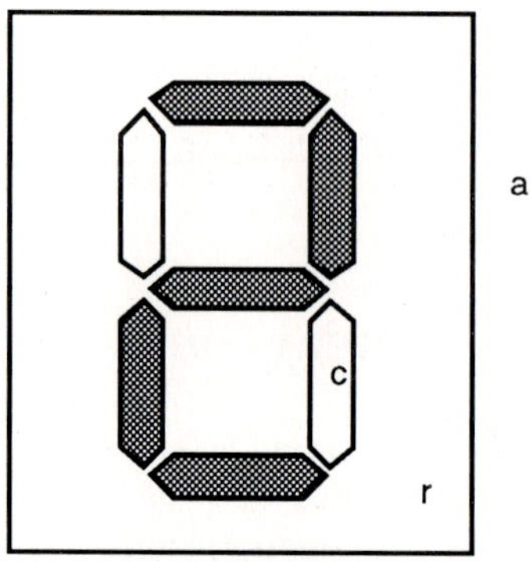

a = anode
c = cathode
r = reference electrode

Fig. 1.1 Alphanumeric character, after reference [65]. The electrodes a, c and r are explained in chapter 2.

Initially ECD development was focused on applications that now employ LCD displays, for example small displays such as watch faces, clocks, radio dials or even personal-computer screens. More ambitiously, television screens and optically addressed computers are envisaged [15]. All these applications require multiple electrodes. For example, a digital watch face uses alpha-numeric characters, each of which comprise seven independent insulated electrodes (Fig. 1.1).

Several ECD applications require only a single 'working' electrode (of at least two – see chap. 2) to produce an expanse of colour. In an optical computer or systems involving optical data storage [15], pixels may represent either 'on' or 'off' when coloured or bleached respectively, and thus interrupting (or not) a beam of light or a laser, but sub-nanosecond response times would be necessary for such purposes, and currently no ECDs are as fast as this. Rates of ECD operation are discussed in chapter 2 (section 2.5).

Electrochromic mirrors [14, 16–20] in cars illustrate another application, discussed further in chapter 3. At night, the lights of following vehicles cause dazzle on reflection from the driver's or the door mirror (Fig. 1.2), which can be prevented by the formation of an optically absorbing electrochrome over the reflecting surface [16]. In such a device, the back electrode is a reflective material enabling the ECD to act as a normal mirror when bleached. Also, when darkened, the electrochromic material must be of only moderate opacity, to allow the mirror to still reflect some light.

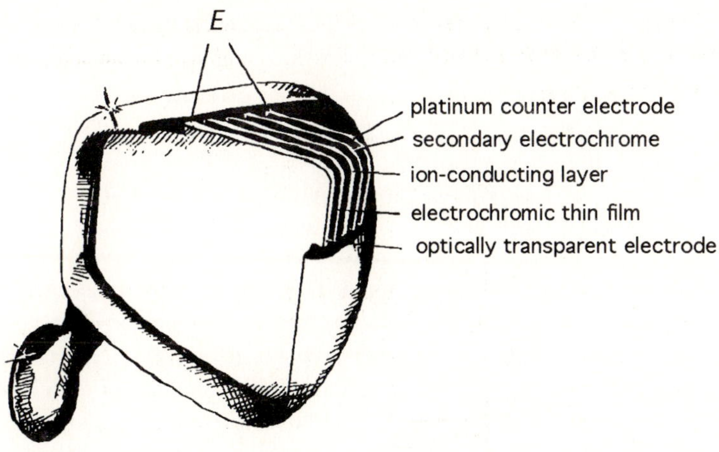

Fig. 1.2 Cutaway diagram of a typical design of a solid-state electrochromic car-door mirror.

Electrochromic sun-glasses have been produced which, unlike photochromic lenses, may be darkened at will. In fact, whole windows may be coloured electrochromically to cut down the light in a room, office or though a car windscreen. Such shutters have been studied extensively by Goldner [21–23]. (The term 'smart glass' was coined by Svensson and Granqvist [24] in 1984; *cf.* 'smart windows', 'smart materials' and similar Americanisms.) Blocking sunlight would require the dissipation of absorbed heat, unless the radiation can be reflected metallically by the electrochrome, implying metallic reflectivity in this material.

1.4 Terminology of Electrochromism

1.4.1 Primary and Secondary Electrochromism

The simplest electrochromic light modulators have two electrodes directly in the path of the light beam. The *primary* electrochromic species is attached to (indeed, part of) the *working electrode*, but there must also be a *counter electrode* (chap. 2), possibly conducting ITO glass. The working electrode could itself be transparent.

If both electrodes bear an electrochromic layer, then the colour formation within the two must operate in a complementary sense, which may be illustrated here with the example of WO_3 and vanadium pentoxide: WO_3 becomes strongly coloured (blue) after being reduced, and effectively colourless when oxidised. By contrast, V_2O_5 is a rich brown/yellow colour when oxidised, yet faintly coloured (blue) when reduced. In an ECD constructed with these two materials, one oxide layer is present in its reduced form while the other is oxidised; thus the operation of the device is:

$$\overbrace{\underset{\text{(pale yellow)}}{WO_3} + \underset{\text{(pale blue)}}{M_xV_2O_5}}^{\text{bleached}} \rightarrow \overbrace{\underset{\text{(blue/metallic)}}{M_xWO_3} + \underset{\text{(brown)}}{V_2O_5}}^{\text{coloured}}$$

M here is a monovalent cation. The tungsten-oxide is termed the *primary* electrochrome since it is the more strongly coloured species and, in this example, V_2O_5 acts as the *secondary* [25]. Secondary electrochromes operate to complement primary electrochromes, one colouring on insertion of counter ions, the other forming colour as such ions are extracted or ions of opposite charge inserted. Clearly, the second electrode need not acquire colour at all. (The fraction x in the solids indicates the fraction of V or of W that has been reduced to respectively the +4 or +5 state.)

In many contemporary investigations, tungsten trioxide is employed as the primary colour-forming species, while the secondary layer is an oxide of iridium [25, 26], nickel [27, 28], niobium [29, 30] or vanadium [31–33], or it could be Prussian blue [34, 35]; in a novel mixed organic–inorganic cell, Dao and Nguyen [36, 37] used poly(*N*-benzyl)-aniline as the secondary electrochrome. Kashiwazaki [38] has used ytterbium *bis*(phthalocyanine) as the primary electrochromic species, with Prussian blue as secondary. Prussian blue is quite intense enough in colour to itself be the primary electrochrome.

1.4.2 Colour and Contrast Ratio

Visible light can be viewed as electromagnetic waves of wavelength 420 nm (violet) to 700 nm (red) or equivalently [39] as particulate photons of energy 4.7×10^{-19} J (violet) to 2.8×10^{-19} J (red). The colours cited refer to light directly entering the eye. However, colour is a subjective visual impression involving retinal responses of the eye to particular wavelengths of the impinging light (table 1.1). Light comprising all visible wavelengths appears white. Reflected colours* result from absorption by the reflecting material of some of these wavelengths, that is, from subtraction from the full wavelength range comprising incident white light. In white light, the perceived colour of a material is the complementary colour of the light it absorbs (Fig. 1.3)† [40, 41].

Table 1.1 Wavelength and Energy Ranges for Perceived Colours of Emitted Light [46] (values given to three significant figures). The numbers above and below each colour represent its range.

	λ/nm	λ^{-1}/cm^{-1}	$10^{-14}\nu$/s^{-1}	$h\nu$/eV	$10^{19}h\nu$/J	$Lh\nu$/kJ mol^{-1}
	750	13,300	4.00	1.65	2.65	159
Red						
	635	15,800	4.72	1.95	3.13	188
Orange						
	596	16,800	5.03	2.08	3.33	200
Yellow						
	580	17,200	5.17	2.14	3.42	206
Green						
	520	19,200	5.77	2.38	3.82	230
Blue						
	470	21,300	6.38	2.64	4.23	255
Indigo						
	440	22,700	6.81	2.82	4.51	272
Violet						
	390	25,600	7.69	3.18	5.09	307
UV						

* The reflection referred to here is more precisely diffuse reflectance [42] which results from reflection by micro-particles of the unabsorbed wavelengths. Specular reflection on the other hand is the almost total reflection of all wavelengths by metal surfaces or polished ('shiny') surfaces generally, as in mirrors.

† Differences between table 1.1 (direct observations of monochromated tungsten emission) and Fig. 1.3 (complementary colours in sunlight) arise partly from the differing white light sources but mostly from compromises attending the approximate notion of complementarity.

The wavelengths (or photon energy) of the absorbed light needs consideration: a 'single' wavelength of absorption is encountered only with single-atom or single-ion photon absorption, the photon energy being transformed into internal electronic energy by the excitation of an electron between precise energy levels associated with the two orbitals accommodating the electron before and after the photon absorption, or 'transition' as it is termed. In molecules the energy levels involved are somewhat broadened by contributory vibrational (and to a lesser extent, rotational) energies. Thus, on light absorption, transitions occur between two 'spreads' of energy levels, (of, however, narrow spread) allowing the absorption of photons with a restricted range of energies, that is, of light of a restricted range of wavelengths, giving an absorption *band*. The maximum absorption, roughly in the centre of such a band, corresponds to the 'average' transition.

The target molecule here is called a *chromophore*, and when the colour resulting from absorption is evoked electrochemically, an *electrochromophore* or more briefly, an *electrochrome* (section 1.1).

The absorption spectrum of a substance represents the relative intensity (relative number of photons) absorbed at each wavelength. It is recorded in a spectrophotometer, in which the sample is illuminated by single-colour (monochromated) light, that is, light of a specific wavelength, steadily changed from 420 nm (violet) to 700 nm (red). The intensity of the transmitted light emerging is monitored by photocell or photomultiplier; much fancier versions of spectrophotometry are available. The spectrum is plotted as absorbance A, or transmittance T in an inverted representation, *vs* wavelength λ (or *vs* λ^{-1}, 'wavenumber', which has the merit of being proportional to the photon energy).

The Beer-Lambert law [43] for optical absorption relates the absorbance, expressed as log of the ratio of the intensities, to the concentration c of chromophore and optical pathlength l through the sample:

$$A = \log\left(\frac{I_o}{I}\right) = \varepsilon\, c\, l \tag{1.1}$$

The proportionality factor ε is the molar extinction coefficient or molar absorptivity of the absorbing species. From the preceding account, it should be clear that ε will vary with wavelength λ since A does, and that it is the parameter quantifying the strength of the optical absorption at each wavelength. $\varepsilon(\lambda)$ (the value at wavelength λ) and ε_{max} (the value at the maximum, often written without subscript) will depend on solvent, or solid matrix, to a greater or lesser extent. When the absorption results from optical CT, Kosower's parameter Z [44], which is the energy (inverse wavelength) for the maximum absorption of a particular chromophore in a given solvent (see Section 8.2), varies with solvent in a manner followed proportionately by other similar chromophores. Z is a useful indicator of solvation in the chromophore–solvent system involved, which will clearly determine the transition energy, that is, where the absorption maximum occurs.

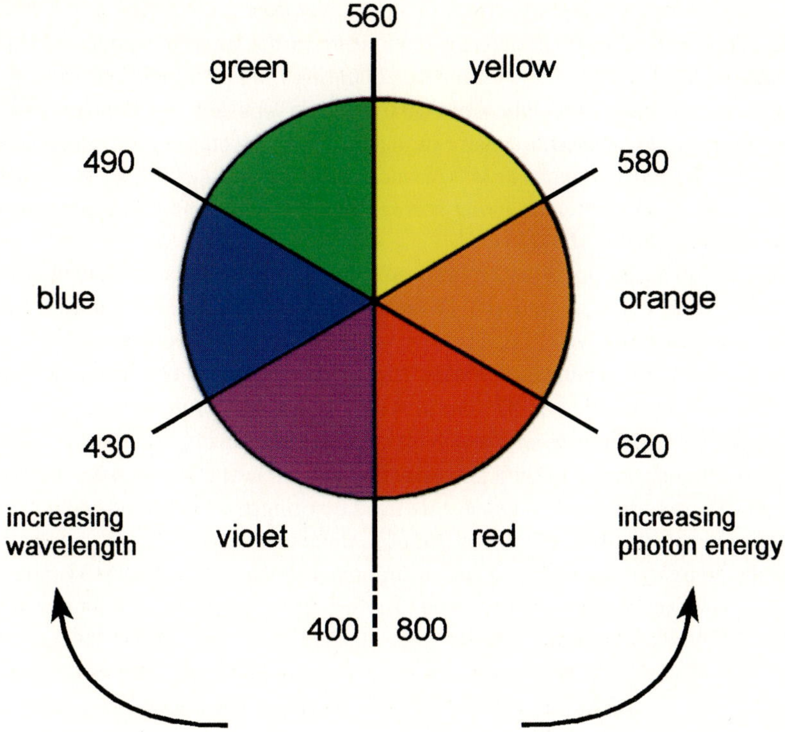

Fig. 1.3 Textbook chart of approximate wavelength ranges (in nm) of reflected colours. Colours in directly opposite segments are called complementary: white light, after absorption (removal) of a particular colour, will show the complementary colour. (The reflected colour observed represents those wavelengths of the incident, polychromatic white light not absorbed by the pigments) [41].

The absorption can thus arise from photo-excitation of an electron from a lower (or ground-state) energy level to a higher one either in the same molecule, which is an *intramolecular excitation*, or within a neighbouring moiety, which involves an intermolecular interaction termed *optical charge transfer*, or 'optical CT'. The redistribution on photon absorption of electron density in the absorbing species is more or less exactly depends on the *transition moment M*. *M* is measured from the area of the absorption band (i.e. of a trace of ε *vs* λ^{-1}); the molar absorptivity at the maximum is commonly taken as being proportional to *M* [42].

The most intense optical absorptions are often a consequence of optical CT, as in $Fe^{3+}CNS^-$ (Fig. 1.4; see section 1.1) since like intramolecular electronic transitions these are processes 'allowed' (favoured) by wave mechanical selection rules for spectral transitions. The permanganate ion MnO_4^- exhibits a deep purple colour characterised by ε = 2,400 $dm^3\ mol^{-1}\ cm^{-1}$ at 525 nm [45] (the wavelength of the maximum of *one* of its bands). Here, electrons from a low-lying orbital predominantly on oxygen are photo-excited to a higher orbital located primarily on the central Mn, in a transition within the anion. If O^{2-} is considered a ligand, this transition might thus fall into *either* class of electronic excitation, but it is best thought of as intramolecular.

To distinguish colouration due to absorption from emitted colour [41] (table 1.1), note that electrons can also be excited by heating, for example in red-hot or white-hot substances, and the subsequent drop from the excited level(s) to a lower level or various lower levels involves the *emission of photons* which are perceived as colours, detailed in table 1.1. Such emission can also result from electrical excitation of electrons as in LEDs, or from preceding photo-excitation, as in fluorescence or phosphorescence.

Finally, the absorption spectroscopy outlined above has to be supplemented for insoluble solids, or solids not otherwise amenable to absorbance measurement, by *diffuse reflectance spectroscopy* [42], in which the absorptions are inferred from diffusely reflected light, monochromated as in absorbance studies. Complications arise from grain-size effects, and the technique is basically less convenient and perhaps less informative than the absorption method.

In any electrochromic system, a quantitative measure of the intensity of the colour change is required. That commonly used is the contrast ratio *CR*:

$$CR = \frac{R_o}{R_x} \tag{1.2}$$

where R_x is the intensity of light diffusely reflected through the coloured state of the display, and R_o is the intensity of light diffusely reflected from the bleached (uncoloured) state from a (diffuse) white back plate [47]. For precision, *CR* should refer to a specific wavelength or relate to an integral value for white light.

The right-hand side of equation (1.2) may be replaced with $\exp(2\alpha l)$ to introduce the linear absorption coefficient* α, and the film thickness l. The factor of two arises because photons must pass through the coloured layer twice.

In transmission mode, the optical absorption of an electrochromic film is related to the injected charge per unit area Q (assuming no side reactions) by an expression akin to the Beer-Lambert law, since Q is proportional to the number of colour centres:

$$A = \log\left(\frac{I_o}{I}\right) = \eta\, Q \tag{1.3}$$

where η is the 'colouration efficiency' of the film (see below). A *CR* of less than 2 or 3 is not easily perceived by eye, and as high a value as possible is desirable. Commonly the *CR* is expressed as a ratio, for example, 7:1, and is best measured at the wavelength of maximum absorption by the coloured state. Equation (1.3) implies a change from zero absorption to the value A.

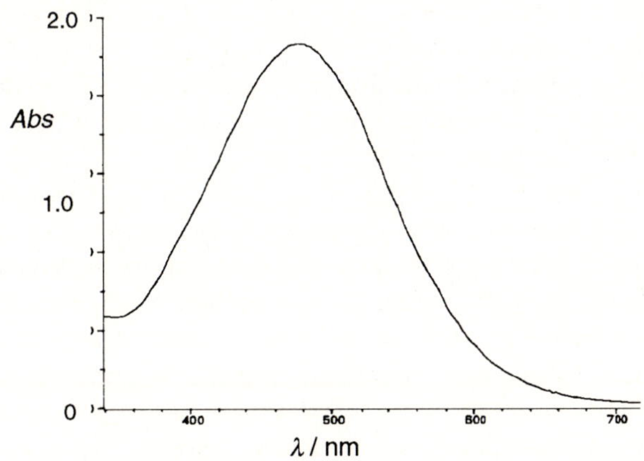

Fig. 1.4 Visible spectrum of the iron(III) thiocyanate charge-transfer complex in water at a concentration of 10^{-3} M in a 1 cm cell.

When there is a great difference in colour between the two redox states, but both are highly coloured (e.g. polypyrrole [48]) then the contrast is not *perceived* to be great. In this case, the *CR* is highly wavelength dependent. If electrochromism is the result of solely a change in oxidation state of a monatomic ion or an atomic species, a low *CR*

* This quantity differs from the molar absorptivity or molar absorption coefficient ε, of the Beer-Lambert law.

value will normally ensue. If, however, optical charge transfer or a similarly allowed internal electronic transition can occur in the product, the *CR* will usually be high since the coloured state then has a large molar absorptivity. Thus a *CR* of 60:1 has been reported for the heptyl viologen system in water [49] where the transition can be viewed as optical CT or an internal transition (see chap. 8).

1.4.3 Colouration Efficiency

The colouration efficiency η is related to an optical absorbance *change* ΔA via equation (1.3), and to the linear absorption coefficient α, film thickness d and charge injected Q per unit area, by the relationship [33]:

$$\eta = \frac{(\alpha\, d)}{Q} = \frac{\Delta A}{Q} \tag{1.4}$$

In the use of these equations, it is assumed that all optical effects are absorptive, that only a single absorbing species is effective at the wavelength chosen for monitoring, and that the Beer-Lambert law is obeyed. η may be regarded as that electrode area which may be coloured to unit absorbance by unit charge. η is (arbitrarily [33]) designated as positive for cathodically induced colouration (by electron gain, or reduction) and negative for anodic colour formation (by electron loss, i.e. oxidation).

If η_p is the colouration efficiency of the primary electrochromophore, and η_s that of the secondary, then the colouration efficiency η_o of the complete ECD device is obtained as $\eta_o = (\eta_p - \eta_s)$. For the most intense electrochromism, the parenthesised entity should be maximised, that is, a large change in A is required from injection of a small charge. This may be achieved in two ways, either by using a large η_p together with an η_s of the opposite sign (i.e. complementary electrochromism), or by using a combination of large η_p and small η_s both of the same sign. In other words, both electrodes colourise simultaneously or unwanted colour in one electrode is feeble.

An example of the latter approach with η values of the same sign is the use of an ITO cathode as the secondary film against WO_3 since ITO does not colour strongly when electro-reduced, becoming only very faint yellow-brown.

In general, organic electrochromes exhibit a greater η than do inorganic species because the molar absorptivities of the former are usually higher. A selection of η values for metal oxide thin films are listed in table 1.2. (Occasional measurements at IR frequencies imply measures of thermal protection rather than colouration.)

Table 1.2 Values of the Colouration Efficiency η for Thin Films of Metal Oxide Electrochrome. (Positive η denotes cathodic colour formation while negative values denote anodic colouration and a dash indicates 'unspecified').

Oxide	State	Preparation[a]	Counter ion	$\eta/cm^2\ C^{-1}$	Ref.
IrO_x	—	—	—	−15 (633)[b]	[51]
IrO_x	—	RF-sputt.	Li^+ or OH^-	−15[c]	[33]
IrO_x	amorph.	anodic dep.	H^+ or OH^-	−30[c]	[52]
MoO_3	amorph.	th.evap.	H^+	77 (700)	[13]
$Mo_{0.008}W_{0.992}O_3$					
	amorph.	th.evap.	H^+	110 (700)	[53]
Nb_2O_5	amorph.	RF-sputt.	Li^+	< 12	[30]
Nb_2O_5	—	—	—	≤ 24[d] (633)	[54]
NiO	amorph.	elec.dep.	H^+ or OH^-	−20	[55]
NiO	—	RF-sputt.	H^+ or OH^-	−36 (640)	[13]
Rh_2O_5	—	anodic dep.	H^+ or OH^-	−20 (546)	[52]
TiO_2	amorph.	RF-sputt.	Li^+	5	[29]
TiO_2	amorph.	—	H^+	8 (546)	[56]
V_2O_5	polycryst.	RF-sputt.	Li^+	−15 (600–1600)	[57]
V_2O_5	polycryst.	RF-sputt.	Li^+	−35 (1300)	[32]
WO_3	amorph.	th.evap.	H^+	115 (633)	[58]
WO_3	amorph.	th.evap.	H^+	76 (550)	[59]
WO_3	amorph.	th.evap.	Li^+	79 (800)	[30]
WO_3	polycryst.	CVD	H^+	38–41	[60]
WO_3	polycryst.	DC-sputt.	Na^+ or Li^+	115 (633)	[61]
WO_3	polycryst.	RF-sputt.	H^+	42 (650)	[62]
WO_3	polycryst.	DC-sputt.	Li^+	109 (1400)	[30]
WO_3	—	—	Li^+	59 (555)	[63]

Key: [a] 'RF-sputt' = RF–sputtering; 'th.evap' = thermal evaporation. 'elec. dep' = electrochemical deposition. 'CVD' = chemical vapour deposition. 'DC–sputt' = DC–magnetron sputtering. [b] Wavelength (λ/nm) used for measurement in parenthesis. [c] Independent of wavelength [52]. [d] Charge passed was calculated from a cyclic voltammogram [54].

1.4.4 Write–erase Efficiency

The write–erase efficiency is the percentage of the originally formed colouration that may be subsequently electro-bleached; it can conveniently be expressed as a ratio of absorbance changes. For a successful display, the efficiency should closely approach 100% – a major test of ECD design and construction. Species remaining in solution in both coloured and uncoloured states, such as methyl viologen, diffuse from the electrode surface after electro-colouration. Since bleaching of such an ECD requires *all* the coloured material to diffuse back to the electrode for electrooxidation, which is a relatively slow process, the write–erase efficiency on a practical time-scale is poor for all-solution systems.

One approach in attaining a high write–erase efficiency is to derivatise an electrode i.e. to chemically tether the electrochrome to the electrode surface. Wrighton [50] has often employed this approach, binding a bipyridilium species to optically transparent electrodes, in the main by use of substituents at nitrogen terminating with groups such as the reactive trimethoxy-silyl moiety in (**II**), that will bond to the oxide lattice on the optically transparent electrode (OTE) face.

A second method of ensuring a high write–erase efficiency is to use a solid (i.e. insoluble) electrochrome that is permanently attached to the electrode, such as Prussian blue, rare-earth phthalocyanines, or iridium oxide and tungsten trioxide films.

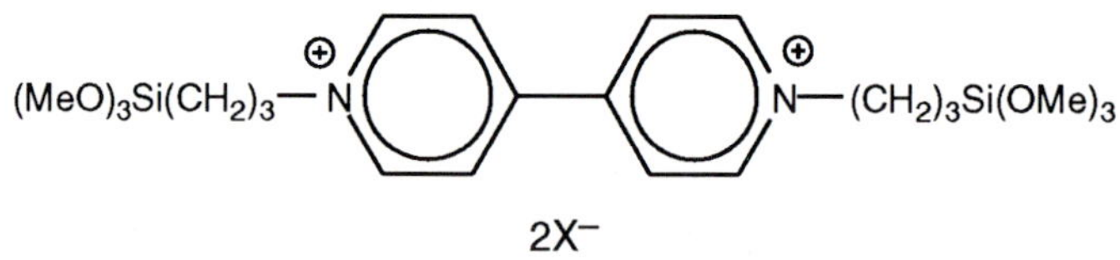

II

A third approach is to have a colourless species in solution which, on electron transfer, precipitates from solution onto the electrode, thereby depositing a coloured film. This occurs with methoxyfluorenes in acetonitrile [64], or long-chain bipyridilium salts in water [65]. An organic electrochrome that has received wide attention is heptyl viologen (HV) dibromide in aqueous solution [65] (species (**I**) of section 1.1, in which $n = 6$). As a dication, HV^{2+} is soluble and only faintly coloured, but forms an intensely coloured red-crimson radical-cation solid film on one-electron reduction to $HV^{+\bullet}$.

A fourth, more recent, approach to improve the write-erase efficiency is the use of solid polymeric electrolytes, for example, poly(ethylene oxide) (PEO), in which the electrochrome is embedded or dispersed, thus slowing its diffusion away from the electrode. This approach has worked well with heptyl viologen [66] (for which it is not really needed) and with methyl viologen or methylene blue, in poly(2-acrylamido-2-methylpropanesulphonic acid), polyAMPS [67]. Solid electrolytes may also be used in ECDs that comprise solid, permanently insoluble, electrochromic layers, for example lutetium *bis*(phthalocyanine) in contact with polyAMPS [68]. Polymeric electrolytes are discussed in more detail in references [69-72].

1.4.5 Response Time

The time required for an ECD to colour from its bleached state (or vice versa) is termed its response time τ. For most devices, τ values are of the order of a few seconds. For ECDs in general, τ is slower than for either LCDs or CRTs, usually because of the necessity for diffusion, either of charged species through the electrode film or, for all-solution systems, of the electrochrome to the electrode. In applications such as electrochromic windows or mirrors, response times of seconds (minutes for windows) can be tolerated, but if devices such as optical switches or television screens are envisaged, then very fast response times will be necessary, as discussed in chapter 2. At present, the fastest displays exhibit response times of the order of 20 ms, claimed for Prussian blue [73], and 10 ms for a solid-state device containing polymeric pyrazoline [74]. The conditions used for determining such rates have not been cited in either case.

Unfortunately, there is no consistency in the criteria employed for determining τ: it may be the time necessary for some fraction (arbitrary or defined) of the colour to form, such as is indicated by a particular increment of optical density, or the time for all or part of the charge to be injected. Because of these inconsistencies in reporting τ, as well as common omission of criteria or conditions employed, any collation of response times would be of severely limited value.

The calculation of electrochemical response time has been treated theoretically for tungsten trioxide ECDs by Green [75, 76] and by Faughnan and Crandall [47] (see chaps. 2 and 4) and these frameworks could be usefully followed in future estimates of τ values.

The use of semi-conducting ITO in the construction of transparent conducting electrodes limits the response time, the slowness arising from its moderate conductivity [77-80]. To decrease the response time, Dove [81] coated a WO_3 film with a thin (semi-transparent porous) layer of gold to act as an additional electrical contact to the electrochromic material. This gave a faster ECD, presumably as a result of enhancement of the electronic conductivity.

1.4.6 Cycle Life

When an ECD is continually cycled between its coloured and bleached states, device failure will eventually occur resulting from physical changes in solid phases or from chemical side reactions – see chapter 3. The *cycle life* is a measure of its stability, being the number of cycles possible before such failure. A major aim of device fabrication is obviously to maximise the cycle life.

In some tests of ECDs, extensive cycling, cited as evidence of robustness, involve cycles of a duration considerably shorter than the response time. Such tests are clearly of extremely limited value. The cycle life is a complicated function of the colouration required in the cycle: the cycle life generally decreases if wide changes in composition are required, that is, if the quantity of charge injected or removed is large. Some authors attempt to address the variation in severity of the test inherent here by denoting a cycle life as involving 'deep' or 'shallow' cycles, cycles of duration markedly less than τ for total colour switching indicating 'shallow' cycles.

Cycle lives, being thus generally ill defined or at best only roughly illustrative, are therefore not tabulated here.

1.4.7 The Insertion Coefficient

For a solid metal oxide AO_n which is the electrochromic material in an ECD, electrode reaction necessitates injection of charge, and counter ions M enter the solid. Where the composition of the reaction product is $M_x AO_n$ the value of x depends on the charge passed and hence on the extent of electrode reaction. The quantity x is termed the *insertion coefficient*. For tungsten trioxide, the insertion coefficient is typically $0 \leq x \leq 0.3$, that is, the compositions of tungsten trioxide and its reduced product fall in the range WO_3 to $M_{0.3}WO_3$.

1.4.8 ECD Appearance

In ECDs the appearance of the electrochromic material is of paramount importance. The electrochromically formed colour should be uniform and even, rather than patchy or streaked. A common cause of patchiness is inhomogeneity of the electrochromic material, which can sometimes be avoided by improved deposition methods. More difficult to control, but more common, is intensity gradation caused by an unevenness of the applied electric field across the surface of the substrate, resulting in more intense colouration at the edges of the electrode. A highly conductive supporting electrode will ensure a more even appearance in this case.

References

[1] J.R. Platt, *J. Chem. Phys.*, 34 (1961) 862.

[2] H. Byker, *Proc. Electrochem. Soc.*, 94–2 (1994) 3.

[3] Y. Oonuk and A. Kondo, *Jpn. Kokai. Tokkyo. Koho. JP* 62,104,891, cited in *Chem. Abstr.* 107: P187,533*u*.

[4] R. Mercier, O. Bohnke, C. Bohnke, G. Robert, B. Carquille and M.F. Mercier, *Mater. Res. Bull.*, 18 (1983) 1.

[5] J.C. Powers, W.R. Heller, J. Kumamoto and W.E. Donath, *J. Am. Chem. Soc.*, 86 (1964) 1004.

[6] N.F. Mott and R.W. Gurney, 'Electronic Processes in Ionic Crystals', Dover, New York, 1964.

[7] D.J. Channon and A. Sussman, in J.I. Pankove (ed.), 'Display Devices', Springer-Verlag, Berlin, 1980. Chap. 4.

[8] J.H. Burroughes, D.D.C. Bradley, A.R. Brown, R.N. Marks, K. Mackay, R.H. Friend, P.L. Burns and A.B. Holmes, *Nature*, 347 (1990) 539.

[9] T.P. Brody and P.R. Malmsburg, *J. Hybrid Microelec.*, II (1979) 29.

[10] A. Yasuda and J. Seto, *Solar Energy Mater.*, 25 (1992) 257.

[11] R.J. Mortimer, *J. Electrochem. Soc.*, 138 (1991) 633.

[12] C.M. Elliott and J.G. Redepenning, *J. Electroanal. Chem.*, 197 (1986) 219.

[13] M. Kikao and S. Yamada, 'Proceedings of the International Seminar on Solid State Ionic Devices', World Publishing Co., Singapore, 1988. p. 359.

[14] H.J. Byker, Gentex Corporation, U.S. Patent No. 4,902,108.

[15] T. Nagamura, Y. Isoda and K. Sakai, *Polym. Int.*, 27 (1992) 125, cited in *Chem. Abstr.* 116 (14) 130670 *t*.

[16] F.G.K. Baucke, *Schott Information,* 1 (1983) 11.

[17] F.G.K. Baucke and J.A. Duffy, *Chem. Brit.*, 21 (1985) 643.

[18] F.G.K. Baucke, *Rivista della Staz. Sper. Vetro,* 6 (1986) 119.

[19] F.G.K. Baucke, *Feinwerktechnik und Meßtechnik,* 94 (1986) 25.

[20] F.G.K. Baucke, *Solar Energy Mater.,* 16 (1987) 67.

[21] R.B. Goldner, F.O. Arntz, G. Berera, T.E. Haas. G. Wei, K.K. Wong and P.C. Yu, *Solid State Ionics*, 53-56 (1992) 617.

[22] R.B. Goldner, *Proceedings of the International Seminar on Solid State Ionic Devices,* World Publishing Co., Singapore, 1988. p. 379

[23] R.B. Goldner, T.E. Haas, G. Seward, K.K. Wong, P. Norton, G. Foley, G. Berera, G. Wie, S. Schutz and R. Chapman, *Solid State Ionics,* 28–30 (1988) 1715.

[24] J.S.E.M. Svensson and C.G. Granqvist, *Proc. S.P.I.E.,* 502 (1984) 30.

[25] S.F. Cogan, T.D. Plante, R.S. McFadden and R.D. Rauh, *Proc. S.P.I.E.,* 823 (1987) 106.

[26] S.F. Cogan, T.D. Plante, R.S. McFadden and R.D. Rauh, *Solar Energy Mater.,* 16 (1987) 371.

[27] S. Passerini, B. Scrosati, A. Gorenstein, A.M. Andersson and C.G. Granqvist, *J. Electrochem. Soc.*, 136 (1989) 3394.

[28] A.M. Andersson, C.G. Granqvist and J.R. Stevens, *Proc. S.P.I.E.,* 1016 (1988) 41.

[29] S.F. Cogan, T.D. Plante, M.A. Parker and R.D. Rauh, *Solar Energy Mater.,* 14 (1986) 185.

[30] S.F. Cogan, T.D. Plante, E.J. Anderson and R.D. Rauh, *Proc. S.P.I.E.*, 562 (1985) 23.

[31] P. Baudry and D. Deroo, *Proc. Electrochem. Soc.*, 90-2 (1990) 274.

[32] S.F. Cogan, N.M. Nguyen, S.J. Perrotti and R.D. Rauh, *J. Appl. Phys.*, 66 (1989) 1333.

[33] S.F. Cogan and R.D. Rauh, *Solid State Ionics*, 28–30 (1988) 1707.

[34] T. Kase, M. Kawai and M. Ura, paper presented at the 1986 S.A.E. Passenger Car Meeting, Dearborn, USA (21–25, Sept., 1986).

[35] N. Kobayashi, M. Nishikawa, H. Ohno, E. Tsuchida and R. Hirohashi, *J. Soc. Photog. Sci. Technol. Jpn.*, 51 (1988) 375.

[36] M.T. Nguyen and L.H. Dao, *J. Electrochem. Soc.*, 136 (1989) 2131.

[37] M.T. Nguyen and L.H. Dao, *Proc. Electrochem. Soc.*, 90–2 (1990) 246.

[38] N. Kashiwazaki, *Solar Energy Mater. Solar Cells*, 25 (1992) 249.

[39] A.I.M. Rae, 'Quantum Mechanics', 2nd Edition, Hilger, Bristol, 1986.

[40] R.W.G. Hunt, 'The Reproduction of Colour', Fountain Press, Tolworth, 1987.

[41] P.W. Atkins and J.A. Beran, 'General Chemistry', 2nd edn., W.H. Freeman, New York, 1992. pp. 231 and 815.

[42] H.-H. Perkampus, 'UV-VIS Spectroscopy and its Applications', Springer-Verlag, Berlin, 1992.

[43] P.W. Atkins, 'Physical Chemistry', 5th edn., Oxford University Press, Oxford, 1994. p. 545.

[44] E.M. Kosower, 'An Introduction to Physical Organic Chemistry', Wiley, New York, 1968.

[45] J.A. Duffy, 'Bonding, Energy Levels and Bands in Inorganic Solids', Longmans, Harlow, 1990.

[46] D.H.G. Crout and D.R. Rosseinsky: direct observation.

[47] B.W. Faughnan and R.S. Crandall, in J.I. Pankove (ed.), 'Display Devices', Springer-Verlag, Berlin, 1980. Chap. 5.

[48] E.M. Genies, G. Bidan and A.F. Diaz, *J. Electroanal. Chem.*, 149 (1983) 101.

[49] D.J. Barclay, B.F. Bowden, A.C. Lowe and J.C. Wood, *Appl. Phys. Lett.*, 42 (1983) 911.

[50] M.S. Wrighton and D.C. Bookbinder, *J. Electrochem. Soc.*, 130 (1983) 1081.

[51] Y. Sato, K. Ono, T. Kobayashi, H. Watanabe and H. Yamanoka, *J. Electrochem. Soc.*, 134 (1987) 570.

[52] W.C. Dautremont-Smith, *Displays*, 3 (1982) 67.

[53] B.W. Faughnan and R.S. Crandall, *Appl. Phys. Lett.*, 31 (1977) 834.

[54] B. Reichman and A.J. Bard, *J. Electrochem. Soc.*, 127 (1980) 241.

[55] C. Liquan, D. Ming, C. Yunfa, S. Chunxiang and X. Rungjian, *Extended Abstracts of the 7th International Conference on Solid State Ionics, Japan*, 1989, abstract 6pB-38.

[56] C.K. Dyer and J.S. Leach, *J. Electrochem. Soc.*, 125 (1978) 23.

[57] S.F. Cogan, N.M. Nguyen, S.J. Perrotti and R.D. Rauh, *Proc. S.P.I.E.*, 1016 (1988) 57.

[58] B.W. Faughnan, R.S. Crandall and P. M. Heyman, *RCA Rev.*, 36 (1975) 177.

[59] I.F. Chang, in A.R. Kmetz and F.K. van Willisen (eds.), 'Non-Emissive Electro-Optic Displays', Plenum Press, New York, 1976.

[60] D. Davazoglou, A. Donnadieu and O. Bohnke, *Solar Energy Mater.*, 16 (1987) 55.

[61] M. Green, W.C. Dautremont-Smith and K.S. Kang, *2nd International Conference on Solid Electrolytes (St. Andrews, Scotland, UK)*, 1978 (cited in ref. [52] above).

[62] M.L. Hitchman, *J. Electroanal. Chem.*, 85 (1977) 135.

[63] K. Matsuhiro and Y. Masuda, *Proc. S.I.D.*, 21 (1980) 101.

[64] B. Grant, N.J. Clecak, M. Oxsen, A. Jaffe and G.S. Keller, *J. Org. Chem.*, 45 (1980) 702.

[65] C.J. Schoot, J.J Ponjée, H.T. van Dam, R.A. van Doorn and P.J. Bolwijn, *Appl. Phys. Lett.*, 23 (1973) 64.

[66] A.F. Sammells, *Government Rep. Announce. Index US*, 87 (1987) Abstr. No. 703,869, cited as *Chem. Abstr.* 107: 86,064*m*.

[67] J.M. Calvert, T.J. Manuccia and R.J. Nowak, *J. Electrochem. Soc.*, 133 (1986) 951.

[68] A.F. Sammells and N.U. Pujare, *J. Electrochem. Soc.*, 133 (1986) 1270.

[69] C.A. Vincent, *Chem. Brit.*, 25 (1989) 391.

[70] P. Hagenmuller and W. van Gool (eds), 'Solid Electrolytes: General Principles, Characterisation, Materials, Applications', Academic Press, London, 1978.

[71] J.R. MacCullum and C.A. Vincent (eds), 'Polymer Electrolyte Reviews – 1', Elsevier, Amsterdam, 1987.

[72] J.R. MacCullum and C.A. Vincent (eds), 'Polymer Electrolyte Reviews – 2', Elsevier, Amsterdam, 1989.

[73] T. Oi, *Ann. Rev. Mater. Sci.*, 16 (1986) 185.

[74] F.B. Kaufman, *Conference Record: Biennial Display Research Conference,* (1978) 23 I.E.E.E. (NY)

[75] M. Green, D.C. Smith and J.A. Weiner, *Thin Solid Films*, 38 (1976) 89.

[76] M. Green, *Thin Solid Films*, 50 (1978) 145.

[77] H. Kaneko and K. Miyake, *Appl. Phys. Lett.*, 49 (1986) 112.

[78] J. Nagai, T. Kamimura and M. Mizuhashi, *Proc. S.P.I.E.*, 562 (1985) 39.

[79] P. Schlotter, *Solar Energy Mater.*, 16 (1987) 39.

[80] O. Bohnke, M. Rezrazi, B. Vuillemin, C. Bohnke, P.A. Gillet and C. Rousselot, *Solar Energy Mater. Solar Cells*, 25 (1992) 361. The effect of substrate conductivity is mentioned on page 366.

[81] A.R. Haranahalli and D.B. Dove, *Appl. Phys. Lett.*, 36 (1980) 791.

2 Electrochromic Systems: Electrochemistry, Kinetics and Mechanism

2.1 Introduction

In this chapter the elements of electrochemistry, which entirely underlie electrochromism, are introduced. Section 2.2 covers fundamental electrochemistry, introducing the use of electrode potentials and their determination in equilibrium conditions within a cell. This section includes the assembly of cells comprising two electrodes. Section 2.3 exemplifies electrochromic operation.

Section 2.4 covers electrochemical methods involving dynamic electrochemistry, particularly cyclic voltammetry, which is important in studying electrochromism; three-electrode systems are required here.

In section 2.5 the rates of mass transport and electron transfer, the two rate-limiting (thus current-limiting) processes encountered during the electrochemistry, are described. Diffusion of both electrochrome and counter ions is discussed to illustrate the way charge-carrier movement limits the rate of redox processes within ECDs. Section 2.6 briefly refers to impedance. Finally, in section 2.7, we discuss the three types of electrochrome used within electrochromic devices (ECDs). More comprehensive treatments of electrochemical theory will be found elsewhere [1–3].

2.2 Equilibrium Electrochemistry

An electrochemical cell comprises at least two electrodes, each made up of two different 'charge states' (more properly, oxidation or redox states) of a particular chemical. These two states stay in equilibrium at only one potential applied to this electrode (comparable statements being also relevant to the other electrode). If the potential applied to the electrode in contact with both redox states is different from this equilibrium potential then one of two 'redox' reactions can occur: electron gain (reduction, eq. 2.1)

$$\mathrm{O} + n\,\mathrm{e}^- \rightarrow \mathrm{R} \tag{2.1}$$

or electron loss (oxidation, the reverse of eq. 2.1). O and R are called a *redox couple* and the potential of the electrode in contact and in equilibrium with the two redox states is the *electrode potential, E* (we shall later re-label this potential E_{oc}). The electrode potential for the O,R couple is related to the ratio of their respective concentrations by a form of the Nernst equation:

$$E_{\mathrm{O,R}} = E^{\mathrm{o}}{}_{\mathrm{O,R}} + \frac{RT}{nF}\ln\left(\frac{[\mathrm{O}]}{[\mathrm{R}]}\right) \tag{2.2}$$

where concentrations are denoted by square brackets, R is the gas constant, F the Faraday constant, T the thermodynamic temperature and n is the number of electrons involved in the electron transfer reaction (eq. 2.1). E^o is the *standard electrode potential,* and is defined as the electrode potential (on a scale defined below) measured at standard pressure and temperature, with both O and R present at unit concentration (or formally and more accurately, at unit activity – a thermodynamic concept [1]). The two oxidation states O and R can be solid, liquid, dissolved or gaseous. A further component required is an (inert) conductive metal if neither O nor R is an electronically conductive solid.

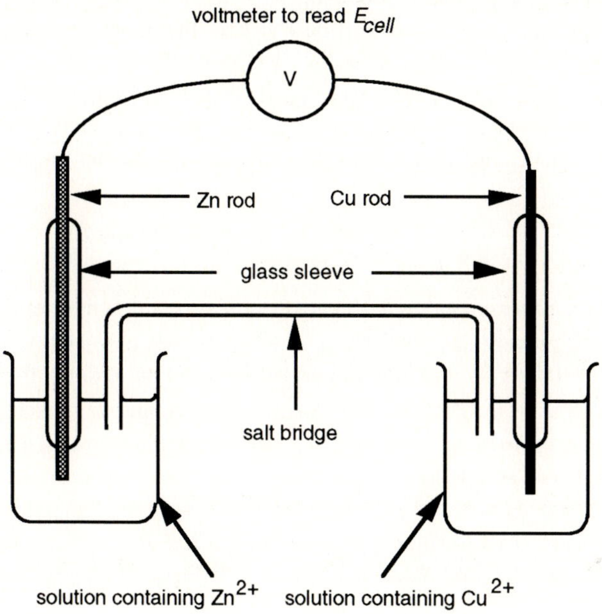

Fig. 2.1 Primitive cell comprising Cu^{2+}, Cu and Zn^{2+}, Zn half cells for equilibrium electrochemical measurement.

Fundamentally, the value of E^o for any particular (O,R) system in eq. 2.1 is determined by the effective condensed-phase electron affinity of O (or, equivalently, the effective condensed-phase ionisation potential of R), on a relative scale fixed by assigning a value to one selected redox system, as below. E for the (O,R) *half cell* cannot be determined independently, since only differences in electric potential may be measured directly. Figure 2.1 shows the simplest electrochemical cell. The lefthand electrode is a zinc rod immersed in an aqueous solution containing Zn^{2+}; these two redox states (Zn^{2+}, Zn) comprise a redox couple. As in Fig. 2.1, one of the redox species in reaction (2.1) also functions as the contact electrode by which E may be monitored, since zinc metal is a good conductor as is copper. The spontaneous reaction is

$$Cu^{2+} + Zn \rightarrow Cu + Zn^{2+} \tag{2.3}$$

and proceeds at each electrode via $Cu^{2+} + 2e^- \rightarrow Cu$ and $Zn \rightarrow Zn^{2+} + 2e^-$, the electrons e^- giving rise to external current flow. Ion motion occurs within the solution phase. Such a cell would therefore spontaneously produce current if the electrodes were connected externally with a conducting wire, whether a negligible current, a moderate one or a large one depending on the *rate* of reaction (2.1) (or its reverse) at the more slowly operating electrode. The direction of the reaction is reflected in the relative values of the two electrode potentials, evaluated as outlined below.

The amount of Zn^{2+} in solution will be constant, that is, at equilibrium, only when the potential applied to the Zn equals the electrode potential $E_{Zn^{2+}, Zn}$ and, simultaneously, the copper redox couple (right hand side of the cell) is only at equilibrium when the potential applied to the copper is $E_{Cu^{2+},Cu}$. Neither electrode potential is known as an absolute or independent value but the difference between the two, E_{cell}, is the measurable quantity:

$$E_{cell} = E_{righthand\ side} - E_{lefthand\ side} = E_{Cu^{2+}, Cu} - E_{Zn^{2+}, Zn} \tag{2.4}$$

E_{cell} is just the observed value of the net electrical potential difference to be *applied* across the cell to prevent any reaction (i.e. to effect zero current flow), or it may be measured on a voltmeter by allowing minuscule current flow through the meter. E_{cell} is the electromotive force (*emf*) of the cell. A *salt bridge* comprising a suitable unreactive salt solution may be used to connect the solutions of the two half cells, to prevent mixing of the half-cell solutions.

Electrodes which do not comprise metal/metal-ion can also be made from an inert metal in contact with two oxidation states of a chemical species dissolved in water or other solvent, or from gaseous, insoluble-salt, or pure-liquid components, as exemplified in what follows.

After measurement of E_{cell}, provided that one of the electrode potentials which comprise E_{cell} is assigned, then the other may determined experimentally. In order to establish a scale, the half cell

$$Pt \mid H_2(g)\ (1\ atm),\ H^+\ (unit\ concentration)$$

is assigned an electrode potential E^o of zero for all temperatures*. (In practice, the Pt is covered by finely divided 'Pt black'.) This is the *standard hydrogen electrode* (SHE), in which the electrode reaction is

---

* For exactness activity should be used in place of concentration [1]. Also, 1 atm pressure = 101.325 kPa. A vertical line represents a phase boundary, here between the Pt metal and $H_2(g)$-saturated solution containing H^+.

$$H^+(aq) + e^- = \tfrac{1}{2} H_2(g) \tag{2.5}$$

It is the standard reference electrode since all *standard electrode potentials* are ultimately cited with respect to it. Note that the *half reactions* (to be thought of as taking place in *half cells*) to which these E^o refer are formally written as *reductions,* that is, with the electron e^- on the left hand side.

A secondary reference electrode almost universally used is the saturated calomel electrode or SCE, $Hg(l)\,|\,Hg_2Cl_2(s)\,|\,KCl$ (aq. saturated), of which the redox reaction is

$$\tfrac{1}{2} Hg_2Cl_2(s) + e^- = Hg(l) + Cl^-(aq) \tag{2.6}$$

Its electrode potential E_{SCE} has the value of 0.242 V on the SHE scale [4]. (Note that S of SCE is 'saturated', referring to the KCl(aq), not 'standard'. Hg_2Cl_2 is but sparingly soluble, as indicated by '(s)' for 'solid', and therefore the chloride solution is effectively saturated with respect to Hg_2Cl_2). A pseudo-reference electrode increasingly employed is a silver wire, but with a quite unknown electrode reaction: presumably Ag_2O on its surface and/or *ad hoc* Ag^+ ions create a reproducible electrode reaction.

While the identification may sometimes seem procrustean, all electrochromic systems can be represented in principle as such electrochemical cells. The redox half-reactions, and cell reactions, represent overall stoichiometries (i.e. are in essence a book-keeping exercise). The detailed steps (adding up to the stoichiometric reaction), by which any reaction proceeds, may differ from the stoichiometric reaction.

2.3 Electrochromic Operation Exemplified

The electron-transfer process during colouration is denoted by the terms *anodic* or *cathodic*: cathodically colouring materials form colour when reduced at an electrode made negative, a *cathode,* and anodically colouring electrochromes are coloured at an *anode,* or positive electrode.

A simple electrochromic example is now considered. Species (**I**) in section 1.1 with $n = 0$ is methyl viologen (MV^{2+}: 1,1'-dimethyl-4,4'-bipyridilium). The standard electrode potential (i.e. *vs* the SHE) for the dication and radical-cation redox states in reaction (2.7):

$$\underset{\text{(colourless)}}{MV^{2+} + e^-_{Pt}} \quad = \quad \underset{\text{(blue)}}{MV^{+\bullet}} \tag{2.7}$$

is $E^o = -0.446$ V [5], a value only *observed* with $MV^{+\bullet}$ present in concentration equal to that of MV^{2+} (see eq. 2.2). Both MV^{2+} and $MV^{+\bullet}$ are completely soluble, so the methyl viologen system is a type 1 electrochrome as defined below. Using the $MV^{2+}/MV^{+\bullet}$ couple, a cell may be constructed and represented schematically (but omitting anions) as

$$\mathrm{Pt} \mid \mathrm{MV^{2+}, MV^{+\bullet}} \mid \mathrm{SCE}$$

The potential of the $MV^{2+}/MV^{+\bullet}$ couple is monitored at the Pt electrode immersed in a solution containing both of the redox states comprising the half-cell; the right hand side of the cell is the saturated calomel electrode. The vertical line $\mid$ is a phase boundary between, on the left, the solid electrode and solution containing MV species and, on the right, between $MV^{2+/+\bullet}$-containing solution and the saturated KCl of the SCE. The voltmeter is of high resistance in order to measure E_{cell} exactly by preventing any appreciable flow of current which would perturb the cell concentrations (and would of itself introduce a further error). For this cell E^{o}_{cell} = 0.242 V – (–0.446 V) = +0.688 V, illustrating the convention $E_{cell} = E_{right} - E_{left}$. Also, the sign *given* to E_{cell} is determined by the *observed* polarity of the right-hand electrode, here positive (as can be readily established). This observation means that to *drive* reaction 2.7 at the $MV^{2+}/MV^{+\cdot}$ electrode, a potential more negative than –0.688 V with respect to the SCE would have to be applied to the Pt electrode; the reverse of the above SCE electrode reaction, reaction 2.6, would then occur in accompaniment, in the cell depicted.

The cell reaction is

$$\mathrm{MV^{+\bullet}} + \frac{1}{2}\mathrm{Hg_2Cl_2(s)} = \mathrm{MV^{2+}} + \mathrm{Hg}(l) + \mathrm{Cl^-} \tag{2.8}$$

where again $MV^{+\bullet} \rightarrow MV^{2+} + e^-$ occurs at its electrode, and the SCE reaction (reaction 2.6) correspondingly at the other electrode. For the cell emf E, the Nernst equation relating the observed potential difference E across the cell with the concentrations in solution of the electroactive species is

$$E = E^{o} - \frac{RT}{F}\ln\left(\frac{[\mathrm{MV^{2+}}]\,[\mathrm{Cl^-}]}{[\mathrm{MV^{+\bullet}}]}\right) \tag{2.9}$$

$$E = E^{o'} - \frac{RT}{F}\ln\left(\frac{[\mathrm{MV^{2+}}]}{[\mathrm{MV^{+\bullet}}]}\right) \tag{2.10}$$

where the constant $[Cl^-]$ term associated with the SCE is included in $E^{o'}$ (here equal to 0.688 V), and terms for pure solids and pure liquids are omitted (being constant they are in fact subsumed within E^{o} and $E^{o'}$). In common electrochromic usage, there would be no $MV^{+\bullet}$ initially present, and Nernst's equation immediately shows that in this case application of a negative potential to the Pt–MV^{2+} electrode substantially less in magnitude than 0.688 V will generate the coloured $MV^{+\bullet}$ species by reaction 2.8 (reversed).

The preceding example perhaps represents an unusual use, even an abuse, of the SCE. An exemplar cell described by Sammells and Pujaru [6] includes heptyl viologen (HV^{2+}:

1,1'-di-*n*-heptyl-4,4'-bipyridilium) as electrochromic material in aqueous solution, initially as soluble colourless dication. A second electrode (more practical here than the SCE which is usually reserved for reference use only) is the cerium(III)/cerium(IV) couple. The cell diagram (from which we have omitted anions and unreactive electrolyte commonly present) with the two reactant solutions separated by membrane, is

$$\mathrm{M} \mid \mathrm{HV^{2+}(aq)}, \{\mathrm{HV^{+\bullet}X^{-}(s)}\} \vdots \{\mathrm{Ce^{4+}(aq)}\}, \mathrm{Ce^{3+}(aq)} \mid \mathrm{M'}$$

where { } indicate the products of cell operation on passage of current, M and M' are the metallic or conductive glass contacts, and the central junction (vertical dots) is an ion-permeable membrane such as Nafion®. Nafion® is a sulphonated fluoropolymer incorporating H^+ or Na^+.

In the absence of a current-generating applied potential between the two electrodes (e.g. with a voltmeter connected across them, or with a potential exactly equal and opposite to E_{cell} applied), the cell is in equilibrium with no current flow. When a sufficient potential, greater than E_{cell}, is applied across the cell (positive connected to the righthand electrode), Ce^{3+} loses an electron

$$\mathrm{Ce^{3+} \rightarrow Ce^{4+} + e^{-}} \tag{2.11}$$

and is replenished by diffusion from the bulk of the solution. Both redox states of cerium are soluble, so the Ce^{4+} diffuses away from the electrode after the electron-transfer reaction there. The two cerium species are similarly coloured and little colour enhancement ensues electrochromically at this electrode.

Now consider the cathode. Heptyl viologen dication accepts an electron

$$\mathrm{HV^{2+} + e^{-} \rightarrow HV^{+\bullet}} \tag{2.12}$$

and is replenished by diffusion from the bulk solution toward the electrode. The resultant radical cation ($HV^{+\bullet}$) is highly coloured hence its use as an electrochrome. In water, $HV^{+\bullet}$ is rapidly precipitated with any anion X^- and forms a solid film of $HV^{+\bullet}\,X^-$ on the electrode. In contrast to the Ce^{4+} of reaction 2.11, there is no diffusion of HV species *from* the electrode since the product is solid, and only counter-ions of unreactive electrolyte, here X^-, already present in solution, move out from the electrode to the bulk of the liquid.

The overall cell reaction driven by passage of current is obtained by combining equations (2.11) and (2.12):

$$\mathrm{Ce^{3+} + HV^{2+} \rightarrow Ce^{4+} + HV^{+\bullet}} \tag{2.13}$$

followed (or accompanied) by precipitation:

$$HV^{+\bullet} + X^- \rightarrow HV^{+\bullet}X^- \text{ (s)} \tag{2.14}$$

No electrons appear in equation (2.13); again, each electrode reaction occurs at its own electrode. The methyl viologen system is similar to heptyl viologen in forming colour, but the radical cation product of the electron transfer is soluble, and so diffuses away from the electrode into the bulk solution. Detailed examination of this group of electrochromes is given in chap. 8, including discussion of the *di*-reduced species.

Equation (2.13) represents a cell reaction in a cell in which current flows when appropriate ranges of potential are applied. Such current will comprise two components, faradaic and non-faradaic. The former current is directly linked with the sum of the electron-transfer reactions effected, and the charge (current-time integrated) indicates directly the extent of the cell reaction, since faradaic current involves that charge which yields product (here, as reaction 2.12), accompanied by 2.11).

Electrochromic operation involves the quantity of electrochrome that changes redox state on passage of current, as governed by Faraday's laws, which are as follows.

1. The number of moles of species formed at an electrode during electrode reaction is proportional to the charge passed;
2. A given charge liberates (or deposits) masses of different species in the ratio of their 'equivalent weights' (relative molar masses divided by the number of electrons involved in the electrode reaction).

One mole of electrons bears a charge of 96,487 C, which is the Faraday constant, F.

Non-faradaic current is caused by processes such as charging of the electric double layer at the electrode–solution interface (a local separation of unreactive-solute ions into layers of anion and cation partly governed by application of potential, which results in excess accumulation of ions of one particular charge sign at interfaces). In precise mechanistic descriptions of electrode processes, double layer effects need to be taken into account; this may be complicated.

2.4 Voltammetry

2.4.1 Introduction to Dynamic Electrochemistry: The Three-Electrode Configuration

In section 2.2, the electrochemistry of equilibrium (zero-current) systems was introduced. In such measurements the prospect of transfer of electrons remains latent and the amounts of electrochrome present do not change, and thus study of the actual electrochromic colouration reaction is precluded. To allow the studies outlined in section 2.3, *dynamic* electrochemistry is used, in which current is passed in a controlled way, by applying a potential E to a particular electrode that is different from the steady value, which we now re-label E_{OC}. The subscript OC means 'open circuit' (implying connection only to a voltmeter); 'zero current' would serve equally well. E_{cell} is equivalently an open-circuit or zero-current value.

Voltammetry is the most common of these dynamic techniques and is useful for discerning rates and mechanism, in addition to the thermodynamic data related to E_{cell} and E^{o} usually obtained at zero current or open circuit. Processes occurring at one electrode are monitored by observing the *current* change when the potential applied 'to' that electrode (actually between that electrode and the 'counter-electrode' – see below) is varied steadily through a range which evokes an electron-transfer process. The current is recorded as a function of the potential impressed on the electrode. The potential is varied steadily, the rate dE/dt being kept constant and known as either the *scan rate* or the *sweep rate, v.*

The electrode at which the electrochemical changes of interest occur is called the *working electrode*, WE. In order that the *potential* at the WE be known, the *potential difference E* between the WE and a third *reference* electrode, RE, is taken. The SCE is usually employed as reference – see section 2.2. The scan rate $\mathrm{d}E/\mathrm{d}t$ is of course also referred to the RE.

In equilibrium (zero-current) electrochemical experiments, voltages between the WE and RE are measured via a voltmeter, and the potential recorded as a function of the externally varied concentration of the electroactive species present, according to the Nernst equation (eq. 2.2). By contrast, during voltammetry, compositions at electrodes are perturbed by the passage of charge at the electrolyte–working electrode interface when electrode reactions occur; while minimal perturbation occurs in electro-analytical experiments, quite substantial changes generally occur in electrochromic processes. If current were to pass through the RE, then the electrode composition would alter and the electrode potential (eq. 2.2) of the RE would change to give inaccurate potential measurement; in addition, the simple passage of current itself shifts the potential of the electrode. For these reasons, no charge can be allowed to flow through the RE. A third electrode, the *counter electrode,* CE, is therefore used to complete the current-flow circuit, to obviate the apparent paradox of requiring current passage at the WE while using a zero-current RE. As the third electrode is needed only because current must flow at the WE, the nature and composition of the counter electrode are largely irrelevant to the operation of a voltammetric cell. The potential of the CE is not monitored although electrode reactions must clearly take place at the CE if current is to flow; if oxidation occurs at the WE then reduction occurs at the CE, and vice versa, hence current. Figure 2.2 illustrates the connections.

The control of voltage across the working-electrode/counter-electrode pair is achieved using a *potentiostat.* This device adjusts the voltage in order to maintain the potential difference across the working and reference electrodes (sensed using a high impedance feed-back loop). The potential is varied in a pre-programmed manner via a function generator. The potentiostat forces current through the working electrode to achieve the potential desired. Microsecond rise times are now available with appropriate instrumentation, and potentials can be controlled to within 1 mV. The current through the reference electrode is typically restricted to a picoamp.

The products of electrode reaction at the CE cannot usually be assumed benign and so the CE is properly excluded from the solution bulk, for example, by placing it in a separate electrolyte-containing compartment of the electrochemical cell, with solution contact via a sinter or frit. No separate solution compartments are needed if an ECD comprises solely solid materials (see section 2.7.3).

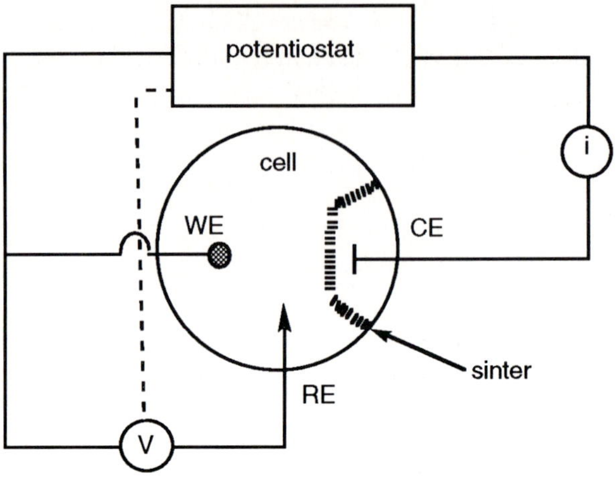

Fig. 2.2 Schematic representation of a three-electrode cell for voltammetric use. Here the voltmeter V is in practice part of the control circuitry of the potentiostat, as is implied by the dotted line.

In voltammetric experiments, an electro-inactive electrolyte like KCl is required in excess of the reactant species, in order to carry current through the cell. Its role at the WE electrode is to effect a kind of charge saturation ('ionic space charge') which forces electroactive species to approach this electrode by diffusion rather than any field-effected conduction process (migration): see also section 2.5.3.

2.4.2 The Use of Voltammetry; Cyclic Voltammetry

Fick's laws (section 2.5.3.2) apply to dynamic electrochemical processes. In voltammetric experiments, electroactive species depleted by electrode reaction are replenished by diffusion from the solution bulk to the surface of the electrode prior to the electron-transfer reaction. Product if soluble diffuses back to the bulk after the electron transfer reaction is complete. The flux of material at the electrode dictates the shape of the current-voltage curve ('voltammogram') obtained. Reactant and product in this example are both dissolved.

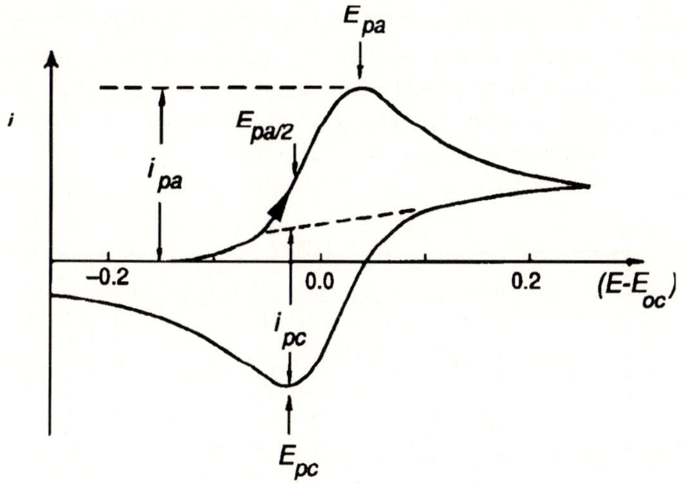

Fig. 2.3 *Typical cyclic voltammogram (CV) obtained for a simple, reversible redox couple in solution, illustrating the key quantities.*

We shall be concerned only with the equation for the peak current i_p as derived by Randles and Sevcik [7, 8]:

$$i_p = 0.4463\, A\, n F \left(\frac{nF}{RT}\right)^{1/2} D_i^{1/2}\, \nu^{1/2}\, c_i \tag{2.15}$$

where A is the area of the electrode and c_i is the bulk concentration of the electroactive species. D_i is its diffusion coefficient (see section 2.5.3). This equation is applicable to 'reversible' all-solution systems (those undergoing rapid electron transfer at the electrode); slow electron transfer results in a different i_p formulation and shape of CV. Furthermore, if reactant or product, or both, are adsorbed or deposited on the electrode, different current-potential curves, following alternative equations, are found. Commonly i_{pc} refers to the peak for a cathodic (reduction) process, while i_{pa} refers to an anodic (oxidative) process.

The peak in the voltammetrically monitored current i_p here is thus proportional to the concentration of the electroactive species, and to the square root of the scan rate. The shape of the curve is governed by the appropriate equations relating to diffusion. In *cyclic* voltammetry, the potential is ramped twice: the potential of the WE starts at an initial value E_{in} and is ramped to a limit, E_λ, known as the *switch potential,* and is then swept back to E_{in}. A plot of current against this potential E is then recorded as a *cyclic voltammogram* or CV. A typical CV for a one electron process is shown in Fig. 2.3. The peak potentials for the anodic and cathodic electron-transfer processes are labelled E_{pa} and E_{pc} respectively. The peak separation is $58/n$ mV for reversible (i.e. rapid) n-electron reactions [1]. Fig. 2.3 also shows the reverse limb of a CV and its corresponding i_p for use in the Randles-Sevcik equation. The same general shape but with peak separation greater than 58/*n* mV indicates 'quasi-reversibility' resulting from somewhat slow electron transfer at the electrode, or from intrusive resistance of electrolyte in a cell comprising less than ideal compositions for voltammetry.

The forward scan rate (E_{in} to E_{λ}) is usually the same as the reverse (E_{λ} to E_{in}). Conventionally the sign of v is cited as a positive number; the range of scan rates commonly employed in cyclic voltammetry is $1 < v < 500$ mV s^{-1}. (The study of ECDs having very small electrodes may allow the use of faster scan rates, up to kV s^{-1}).

A measure of the experimental *time-scale t* is the time required for one voltammetry cycle, that is, for $E_{in} \rightarrow E_{\lambda} \rightarrow E_{in}$: $t = 2(E_{in} - E_{\lambda})/v$. The reversibility (shown as reproducibility or superimposability) or otherwise of successive cyclic voltammograms can give indirect information about the fate of species formed during the forward ramp. The variation of scan rate v provides additional information. For example, while $i_p \propto v^{1/2}$ governs all-solution systems, i_p depends linearly on v for adsorbed systems on the WE: the latter v dependence is given by the appropriate equations for i_p derived from the Langmuir adsorption isotherm.

The combination in spectroelectrochemistry of voltammetry on a transparent electrode with simultaneous absorbance spectrophotometry provides fundamental insights. A novel reflectance method [27] associated with voltammetry is given in section 8.3.3.4.

2.5 Charge Transfer and Charge Transport

2.5.1 The Kinetics of Electron Transfer

The rate of electron transfer at an electrode is a function of the gradient of electric potential applied to the electrode, and follows the Butler-Volmer equation [1, 2] which, for a reduction reaction $O + n\,e^- \rightleftharpoons R$ is

$$i = nFA\,k_f\,c_R(\exp(-\alpha_f\, n\, \phi\, \eta)) - nFA\,k_b\,c_O\,(\exp(\alpha_b\, n\, \phi\, \eta)) \qquad (2.16)$$

where, for brevity, $\phi = F/RT$; c_O is the concentration of the oxidised form of the electro-active species (starting material) and c_R that of the reduced form; α is a fraction termed the transfer coefficient (subscripted *f* and *b* for forward and back reaction respectively), itself a measure of the symmetry of the energy barrier to the electron transfer [1]; η is the overpotential, $(E - E_{oc})$, where E is the potential applied to the electrode and E_{oc} is the zero-current electrode potential. k_f and k_b are the rate constants of electron transfer for the forward and back processes.

To simplify, consider a system with $\alpha_b = (1 - \alpha_f)$ and $c_O = c_R = c$, which makes $k_f = k_b$ which are now represented by k_s. When the rates of forward and back electrode reaction are equal, that is, when no *net* current flows, η then is zero and the quantity $(n\,FA\,k_s\,c)$ is termed the standard exchange current i_o and is a measure of the rate of electron transfer at the electrode. For WO_3, i_o has been measured as 5–30 μA cm^{-2} in

1.0 M H_2SO_4 [9], 80–130 μA cm^{-2} in 0.75 M $LiAsF_6$ in propylene carbonate [10] and 2000 μA cm^{-2} in 1.0 M $LiClO_4$ in propylene carbonate [11]. The cause of these large variations has not been established in detail but differences in the solution compositions, method of sample preparation or internal moisture content clearly can, and do, have a marked effect.

2.5.2 The Use of Semiconducting Electrodes

In the construction of electrochromic display devices (chap. 3), the substrate most commonly used as the optically transparent electrode (OTE) is indium tin oxide (ITO) as a thin film on glass. The thickness of the ITO layer is typically 0.3 μm. Indium(III) oxide, when doped with *ca.* 8 % tin(IV) oxide, is a semiconductor of conductivity *ca.* 8×10^{-4} $S\,cm^{-1}$ [12], so the thickness and exact conductivity of the ITO layer will affect the ECD response time (see section 1.4.5). The rate of supply (flux j) of electrons through the conductor and corresponding current i are obtained from the generally applicable equations:

$$i = n\,A\,\upsilon\,e \qquad \text{and} \qquad j = n\,\upsilon \tag{2.17}$$

where n is the number density of charge carriers, A the cross-sectional area of the conductor, υ the electronic velocity and e the electronic charge. In thin-film indium tin oxide (ITO), the number of charge carriers is relatively small, restricting the rate of charge uptake or loss at the ITO/electrochrome interface. This paucity of transferrable charge is often referred to as the effect of 'poor conductivity' within the ITO. However, some authors believe that the response time of an ECD device with very thin films of ITO depends rather on the *rate* of electron transport, that is, υ, through the ITO [14, 15].

2.5.3 The Rate of Mass Transport

Before the electron transfer reaction can occur, of necessity material must move from the solution bulk and approach close to an electrode. This movement is 'mass transport', and proceeds via three separate mechanisms: migration, convection and diffusion. Mass transport is formally defined as the flux j_i of electroactive species i to an electrode, as defined in the Nernst-Planck equation:

$$j_i = \underbrace{\mu_i\,c_i\left(\frac{\partial \phi}{\partial x}\right)}_{\text{migration}} + \underbrace{c_i\,\overline{\upsilon}_i}_{\text{convection}} - \underbrace{D_i\left(\frac{\partial c_i}{\partial x}\right)}_{\text{diffusion}} \tag{2.18}$$

where μ_i is the ionic mobility of the species i; ϕ is the strength of the electric field, $\overline{\upsilon}_i$ is the (vectorial) velocity of solution (where applicable), and D_i and c_i are respectively the diffusion coefficient and concentration of species *i*. Convection will not concern us further since it is irrelevant for solid electrolytes and otherwise uncontrolled in other ECDs.

2.5.3.1 Migration

Migration is the movement of ions through solution or solid in response to an electric field, a (positive) anode attracting any negatively charged anions, the cathode attracting the cations. For liquid electrolytes containing an excess of unreactive ionic salt, migration may be neglected since the transport number of (i.e. fraction of Ohm's-law current borne by) the electroactive material becomes negligibly small. Migration is an important form of mass transport for ionic movement within solid polymer electrolytes or solid-solution electrochromic layers since transport numbers of the electroactive species become appreciable [15]. The phenomenon of electrode 'polarisation' by excess unreactive electrolyte (the build up of concentration of oppositely-charged electrolyte ions at an electrode) brings about diminution then suppression of migration; diffusion is then the only remaining means of approach to the electrode available to a possibly electroactive species, as follows.

2.5.3.2 Diffusion

Of particular interest to any kinetic study is the diffusion coefficient D of the diffusing species, being representative of its spontaneous motion. Diffusive behaviour obeys Fick's laws [1, 2], the first being for the flux j_i,

$$j_i = -D_i \left(\frac{\partial c_i}{\partial x}\right) \tag{2.19}$$

where $(\partial c_i/\partial x)$ is the concentration gradient, the change in concentration of species i per unit distance. In electrochemical processes, $(\partial c_i/\partial x)$ arises (i.e. is non-zero) because some of the electroactive species is consumed around the electrode; diffusion is evoked by the subsequent concentration gradient.

Fick's second law describes the time dependence of diffusion:

$$\left(\frac{\partial c_i}{\partial t}\right) = D_i \left(\frac{\partial^2 c_i}{\partial x^2}\right) \tag{2.20}$$

The required integration of the second order differential equations often leads to difficulty in the accurate modelling of diffusive systems. However, a useful approximate solution to Fick's second law gives

$$l \approx \sqrt{D_i\, t} \qquad (2.21)$$

where D_i is the diffusion coefficient of species *i*, and *t* is the time required for species *i* to move a distance *l*.

Suppose an electrochrome is deposited on a conductive substrate to a thickness of one monolayer (*ca.* 5×10^{-8} cm thick) and say the solution phase diffusion coefficient of a counter ion is 10^{-6} cm^2 s^{-1}. The response time necessary for the movement of counter ions in to the electrochrome is then $l^2/D \approx 2.5$ ns. A monolayer of electrochrome would be insufficient to generate a satisfactory change in optical absorbance. If sufficient electrochrome were present on an electrode for absorbance changes to be perceptible, say 10 monolayers, then counter ions would need to diffuse *through* the layer of solid electrochrome. The diffusion coefficient through such a solid will always be smaller than through solution. If *D* is say 10^{-9} cm^2 s^{-1}, from eq. (2.21) with $l = 5 \times 10^{-7}$ cm, the response time τ now becomes 0.25 ms, which is still satisfactorily fast. Clearly, the requirement for the optical absorbances to be higher necessitates slower response times τ.

Another indicator of the rate of ionic movement is the ionic mobility μ (velocity υ divided by the driving field), which is related to the diffusion coefficient *D* by the Nernst-Einstein equation

$$\frac{D}{\mu} = \frac{k_B T}{z\, e} \qquad (2.22)$$

where symbols have their usual electrochemical meanings and k_B is the Boltzmann constant. It should be noted that eq. (2.22) cannot be used to compare values of *D* with values of $(k_B T/z\, e) \times \mu$ (from conductivity experiments) if the electrolyte is a solid polymer [15]. Furthermore, in a low-permittivity organic solvent, owing to considerable ion association resulting in the formation of ion pairs and triplets, measured ('apparent') values of *D* show a strong dependence on the nature of the surrounding electrolyte, requiring resolution before *D* can be related to μ values from conductimetry [38].

When the impressed potential of an electrode is stepped from a value giving zero current to one at which the current can reach a maximum, then all the electroactive material at the electrode–solution interface will undergo electrochemical change 'instantly' (reaction 2.1). Electroactive material from the solution bulk then diffuses toward the electrode, coming (diffusing) from successively further distances from the electrode. The flux at the electrode therefore decreases with time. The Cottrell equation [8] describes this current/time response as

$$i = n\, F A \sqrt{\frac{D}{\pi\, t}} \qquad (2.23)$$

The Cottrell equation is a convenient means of determining diffusion coefficients of solution-phase species [1]. The fall-away of current somewhat beyond E_p, as in the CV in Fig. 2.3, is governed by this equation.

2.6 AC or RF Electrochemistry: Impedance or Complex Permittivity Studies

A further electrochemical technique used increasingly is AC or RF impedance [1–3]. In a direct current (d.c.) experiment, application of a voltage V across a resistance R induces current i according to Ohm's law,

$$V = i\,R \tag{2.24}$$

Similarly, application of a sinusoidally varying potential to an electrochemical cell (perhaps superimposed on a pre-set or 'offset' voltage) induces an alternating current (AC) [2]. The AC analogue of Ohm's law is

$$\overline{V} = \overline{i}\,Z \tag{2.25}$$

where Z is the *impedance* and the bars imply time-dependent quantities.

The impedance Z comprises two components, real and imaginary (Z' and Z'' respectively)

$$Z = Z' - \mathrm{j}\,Z'' \tag{2.26}$$

where $\mathrm{j} = \sqrt{-1}$; the term *complex AC* impedance involves tautology. Plots of Z' *vs* Z'' form an Argand plane, termed Nyquist plots in the electrochemical literature.

The impedance of a cell may be measured over a wide frequency range (typically 10^6–10^{-2} Hz) where 10^{-2} to 10^5 Hz comprises the AC range and 10^6 – 10^8 is the radio-frequency 'RF' range. The components Z' and Z'' are determined from the time lag experienced between current $\overline{i}$ and voltage $\overline{V}$, to give a Nyquist plot. The frequencies of features such as maxima in the Nyquist plot relate to the time scales of corresponding processes in the cell. Diffusion coefficients are also obtained.

It is a common aim to mimic the AC response of a cell or electrochemical system using simple combinations of resistors and capacitors which thus form an *equivalent circuit*. Construction of such a circuit is useful evidence for elucidating the processes occurring in electrochemical systems. A specialist monograph gives details [16]. Alternative ways of viewing such measurements involve their presentation as complex permittivity (dielectric relaxation), and also as admittance.

2.7 Electrodes: Classification of Electrochrome Type

The type of electrochrome used in an ECD governs the kinetic behaviour evinced. A classification is presented here, borrowed largely from an earlier treatment by Chang *et al.* [17]. The simplest case is type 1 which comprises species permanently in solution. Type 2 electrochromes are initially present in solution but the coloured product of the electron-transfer reaction is solid. Type 3 materials are always solid attached to the electrodes as thin films. The characteristics of these three types will be discussed in turn, illustrating the physicochemical processes which govern the rate of colour formation.

2.7.1 Type 1 Electrochromes: Always in Solution

For type 1 electrochromes, the electrochemical electron-transfer reaction occurs at the solid–liquid interface. At reaction onset, the electroactive species in solution encounters the solid electrode, undergoes an electron-transfer reaction, be it reduction at a cathode or anodic oxidation, and then in its new form moves away from the electrode, returning to the solution bulk. The electrochrome changes colour on undergoing the electron-transfer reaction. A simple example of a type 1 electrochrome is the methyl viologen dication (above) in water. MV^{2+} in the initial solution is colourless (or faint yellow in the presence of some anions), a bright blue radical cation $MV^{+\bullet}$ being formed when the electrode is made cathodic (negative). The rate-limiting process during the (reductive) electron-transfer reaction is the rate at which MV^{2+} dication reaches (diffuses toward) the electrode from the solution bulk, the rate of electron acquisition when MV^{2+} reaches the electrode interface being much faster; see section 2.5.

In electrochromic devices based on methyl viologen, if a fixed potential, substantially more negative than the (zero-current) electrode potential for MV^{2+} reduction, is applied to the working electrode, then the observed current follows the Cottrell [18] time dependence, $i \propto t^{-1/2}$. This dependence follows from the laws of diffusion: section 2.5.3.2. Hence the absorbance $A \propto t^{+1/2}$ if the current flow is wholly faradaic, that is, if each electron transferred generates a colour centre†. (The t exponent follows because the amount of $MV^{+\bullet}$ generated depends on the integral of i with time.)

Other type 1 electrochromes include transition metal ions and complexes, for example, Fe(III) thiocyanate in aqueous media and hexacyanoferrates (section 1.1), and quinones, TCNQ or TTF in acetonitrile solution.

† There is the complication with viologen species that in aqueous solution, radical cations generally form differently coloured (diamagnetic) dimers $(MV^{+})_2$ [19]. For $MV^{+\bullet}\ I^-$ solution, the equilibrium constant for dimerisation $K_{dim} = 380\ M^{-1}$ [20].

2.7.2 Type 2 Electrochromes: Solution-to-Solid

Type 2 electrochromes are initially soluble and colourless but, following electron transfer, they form a coloured solid film on the surface of the electrode. An example of a type 2 electrochrome is heptyl viologen (HV) referred to in section 2.3 in examples of cells. HV^{2+} dication dibromide is pale yellow in aqueous solution but forms a layer of deeply coloured radical cation salt on reduction. The reduction is, in fact, a two-step process (possibly concerted [21]) involving first an electron transfer reaction

$$HV^{2+} + e^- \rightarrow HV^{+\bullet} \tag{2.12}$$

and subsequent precipitation

$$HV^{+\bullet} + X^- \rightarrow HV^{+\bullet}X^- \text{ (s)} \tag{2.14}$$

The deposit is initially amorphous but becomes more crystalline soon after formation [22]. The morphology, solubility and colour of such radical-cation salts depend on the accompanying counter-anion [22].

The Cottrell current-time relationship of $i \propto t^{-1/2}$ together with $A \propto t^{+1/2}$ is followed fairly closely during short times as soluble HV^{2+} diffuses toward the electrode [23]. At longer times, however, the kinetic response becomes more complicated: $HV^{+\bullet}X^-$ is a solid dielectric of relatively poor conductivity, and further colouration involving generation of additional coloured product must proceed via slow electron transport through the solid deposit at the electrode/solution interface. Again, during electro-oxidative colour erasure of the radical cation salt achieved by the reversal of reactions (2.12) + (2.14), complications in this system become apparent, which necessitate the presence in the electrochemical deposition solution of electron mediators such as ferrocyanide ion which act catalytically to facilitate electron transfer [24, 25].

Other examples of type 2 materials are benzyl paraquat [26, 27], aqueous cyanophenyl paraquat [28] or polymethoxyfluorene species in acetonitrile solution [29]. The deposition of a metal ion (section 2.2) onto an uncoloured or slightly coloured electrode substrate, resulting in pronounced darkening, is a viable type 2 system only rarely employed in electrochromism.

2.7.3 Type 3 Electrochromes: All-Solid Systems

The electrochromes most commonly employed for display purposes are permanently insoluble thin films, and are called type 3 here. Rare-earth phthalocyanines (chap. 5), Prussian blue (chap. 6), and metal oxides, for example, of tungsten, molybdenum, nickel or vanadium (chap. 4), have all been used to this end. For metal oxides and Prussian blue, the electrogenerated colour appears as a consequence of an optical charge-transfer transition between metal centres in the solid-state lattice (see introduction to chap. 4). For a fundamental account of solid-state properties, see texts by Mott [30] and Cox [31].

The electrochemistry of rare-earth phthalocyanine compounds is generally focused on the ligand; electron transfer to the metal can result in decomposition [32]. For electroneutrality, counter ions are injected or inserted into the solid host film in the formation of a solid solution comprising both redox species, or (less commonly in EC systems) of a mixture of the redox systems.

Table 2.1 Diffusion Coefficients D of Various Electrochromic Species. Diffusion coefficients for types 1 and 2 refer to movement of electrochrome, but for type 3 refer to movement of counter-ion species through the solid.

Compound	Ion : Solvent	D/cm^2 s^{-1}	Ref.
Type 1			
Methyl viologen[a]	Cl^- : H_2O	8.6×10^{-6}	[33]
Type 2			
Cyanophenyl paraquat[a]	BF_4^- : PC	2.1×10^{-6}	[34]
Type 3			
Cerium(IV) oxide[b]	Li^+ : PC	5.2×10^{-13}	[35]
Lutetium *bis*(phthalocyanine)[b]	Cl^- : H_2O	10^{-7}, [c]	[36]
Nickel hydroxide[b]	H^+ : H_2O	2×10^{-7} to 2×10^{-9}	[37,38]
Poly(isothianaphthene)[b]	BF_4^- : PC	10^{-14} [b]	[39]
Tungsten(VI) Trioxide[b, d]	H^+ : HCl (aq)	2×10^{-8}, [e]	[40]
Tungsten(VI) Trioxide[b, f]	Li^+: PC	2.1×10^{-11} [g]	[35]
Vanadium(V) Trioxide[b]	Li^+ : PC	3.9×10^{-11}	[35]

Key: PC = Propylene carbonate. [a] Diffusion coefficient of electrochrome diffusing in solution. [b] Permanent film; diffusion coefficient of ion *through* the electrochromic layer. [c] D was apparently calculated from chloride ion mobility. [d] Thermally-evaporated sample. [e] Chronoamperometric measurement. [f] Sputtered film. [g] D from impedance measurement. (Additional values of D for cationic movement through solid WO_3 are given in table 4.1).

The rate-limiting process usually encountered with type 3 electrochromes is the ionic charge transport through the solid electrochrome. The diffusing species, say H^+, enters the electrochrome via the electrolyte–film interface and then moves through the film (see section 2.5.3 on mass transport) although it does not reach the metal-conductor substrate at the other side of the electrochrome. Indeed, an undesirable side-reaction ('spillover') would occur were it to do so, producing gaseous hydrogen. Rather, electron transfer occurs at immobilised electrochrome centres, the protonic charge entering the solid as counter ion in order to maintain electroneutrality. Its subsequent bonding is important.

Two separate diffusion processes will occur within the solid: both the ion diffusion from the electrolyte/solid interface, and electron diffusion from the electrode–electrochrome interface. *Only the slowest, rate limiting, of these diffusion processes will be measurable* in systems with fast electron transfer in the redox process. Both diffusion processes occur during the bleaching of viologen radical-cation deposits [16], colouration of lutetium *bis*(phthalocyanine) [36] or during the electrochemical oxidation of conducting polymers such as polypyrrole (e.g. reference [41]). In all these cases, it is the ionic diffusion that is rate limiting.

In the rate equations in section 2.5, only the slowest diffusion coefficient within the film will be discernible, and is termed the *chemical diffusion coefficient*. Also note that so-called ohmic migration (charge motion driven directly by the electric field) will be evident in the film (see section 2.5.3.1). Table 2.1 contains some typical diffusion coefficients D for ionic motion, both of electrochromes in solution and of counter ions through *solid-state* systems.

References

[1] A.J. Bard and L.R. Faulkner, 'Electrochemical Methods', Wiley, New York, 1980.

[2] The Southampton Electrochemistry Group, 'Instrumental Methods in Electrochemistry', Ellis Horwood, Chichester, 1985.

[3] P.A. Christensen and A. Hamnett, 'Techniques and Mechanisms in Electrochemistry', Blackie, Glasgow, 1994.

[4] D.I. Hitchman and A.C. Taylor, *J. Am. Chem. Soc.*, 59 (1937) 1813.

[5] C. Bird and A.T. Kuhn, *Chem. Soc. Rev.*, 10 (1981) 49.

[6] A.F. Sammells and N.U. Pujaru, *J. Electrochem. Soc.*, 133 (1986) 1270.

[7] J.E.B. Randles, *Trans. Faraday Soc.*, 44 (1948) 327.

[8] A. Sevcik, *Coll. Czech. Chem. Commun.*, 13 (1948) 349.

[9] B.W. Faughnan and R.S. Crandall, *Appl. Phys. Lett.*, 28 (1976) 95.

[10] C.K. Ho, I.D. Raistrick and R.A. Huggins, *J. Electrochem. Soc.*, 127 (1980) 343.

[11] S.K. Mohapatra, *J. Electrochem. Soc.*, 125 (1978) 284.

[12] C. Bohnke and O. Bohnke, *J. Appl. Electrochem.*, 18 (1988) 715.

[13] J. Nagai, T. Kamimura and M. Mizuhashi, *Proc. S.P.I.E.*, 562 (1985) 39.

[14] H. Kaneko and K. Miyake, *Appl. Phys. Lett.*, 49 (1986) 112.

[15] P.G. Bruce, in 'Polymer Electrolyte Reviews – I', (eds.) J.R. MacCullum and C.A. Vincent, Elsevier, Amsterdam, 1987. p. 237.

[16] J.R. MacDonald (ed.), 'Impedance Spectroscopy – Emphasising Solid Materials and Systems', Wiley, New York, 1987.
[17] I.F. Chang, B.L. Gilbert and T.I. Sun, *J. Electrochem. Soc.*, 122 (1975) 955.
[18] F.G. Cottrell, *Z. Physik. Chem.*, 42 (1902) 385.
[19] A.J. Bard, A. Ledwith and H.J. Shine, *Adv. Phys. Org. Chem.*, 13 (1976) 155.
[20] E. Kosower and J.L. Cotter, *J. Am. Chem. Soc.*, 86 (1964) 5524.
[21] R.G. Barradas, S. Fletcher and L. Duff, *J. Electroanal. Chem.*, 100 (1979) 759.
[22] J. Jasinski, *J. Electrochem. Soc.*, 124 (1977) 637.
[23] J. Bruinink and P. van Zanten, *J. Electrochem. Soc.*, 124 (1977) 1232.
[24] J.G. Kenworthy, ICI Ltd., British Patent: 1,314,049 (1973).
[25] A. Yasuda, H. Mori and A. Ohkoshi, *J. App. Electrochem.*, 14 (1984) 323.
[26] B. Scharifker and C. Wehrmann, *J. Electroanal. Chem.*, 185 (1985) 93.
[27] D.R. Rosseinsky, J.D. Slocombe, A.M. Soutar, P.M.S. Monk and A. Glidle, *J. Electroanal. Chem.*, 258 (1989) 233.
[28] D.R. Rosseinsky, P.M.S. Monk and R.A. Hann, *Electrochim. Acta*, 35 (1990) 1113.
[29] B. Grant, N. Clecak, M. Oxsen, A. Jaffe and G. Kellar, *J. Org. Chem.*, 45 (1980) 702.
[30] N.F. Mott, 'Conduction in Non-Crystalline Materials', 2nd Edn., Clarendon Press, Oxford, 1993.
[31] P.A. Cox, 'The Electronic Structure and Chemistry of Solids', Oxford University Press, Oxford, 1987.
[32] J.T.S. Irvine, B.R. Eggins and J. Grimshaw, *J. Electroanal. Chem.*, 271 (1989) 161, and references therein.
[33] T. Kuwana and N. Winograd, *J. Am. Chem. Soc.*, 92 (1970) 224.
[34] D.R. Rosseinsky and P.M.S. Monk, *J.C.S., Faraday Trans.*, 89 (1992) 219.
[35] P. Baudry, thése nouveau regime, I.N.P.G., Grenoble, France, 1989.
[36] M.M. Nicholson and F. Pizzarello, *J. Electrochem. Soc.*, 127 (1980) 821.
[37] D.M. MacArthur, *J. Electrochem. Soc.*, 117 (1970) 729.
[38] C.J. Slaidin and P.D. Lutovtsev, *Electrochim. Acta*, 6 (1962) 17.
[39] H. Yashima, M. Kobayashi, K.-B. Lee, D. Chung, A.J. Heeger and F. Wudl, *J. Electrochem. Soc.*, 134 (1987) 46.
[40] B.W. Faughnan, R.S. Crandall and M.A. Lampert, *Appl. Phys. Lett.*, 27 (1975) 275.
[41] K. Aoki and Y. Tezuka, *J. Electroanal. Chem.*, 267 (1989) 55.

3 Construction of Electrochromic Devices

3.1 Introduction

Many ECDs comprise a sandwich structure of thin layers, the number and nature of which depend on the intended use. The construction of ECDs will be the same whether a large area cell or one small element ('pixel') of a multi-electrode array is required [1].

Electrochromism is used in one of two modes: in adjustments of either reflected or transmitted intensity, Fig. 3.1. These modes will be treated in turn in sections 3.2 and 3.3. Electrolytes are treated in sections 3.4 and 3.6. Methods for preparing thin solid films of electrochromic material are briefly surveyed in section 3.5. Finally, section 3.7 describes the operation of a commercially-available electrochromic rearview mirror.

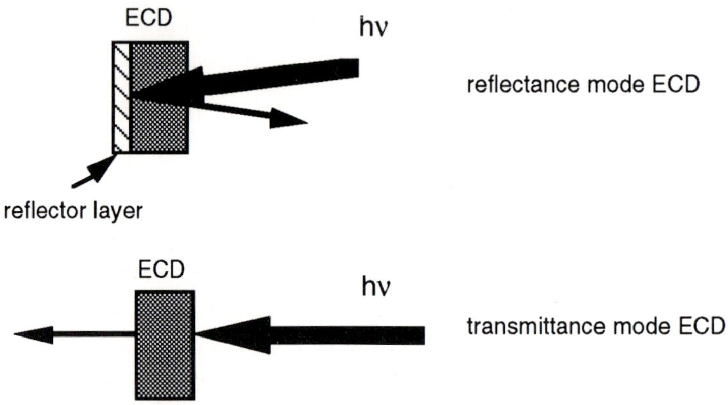

Fig. 3.1 Schematic diagram showing the different modes of ECD operation: (a) reflectance mode and (b) transmittance mode.

3.2 All-Solid Cells with Reflective Operation

For all-solid systems, reflective cells may be assembled according to the schematic diagram in Fig. 3.2. The front panel is an optically transparent electrode (OTE), that is, a solid support, for example, glass bearing a thin transparent but conductive film on its solution-facing side. Tin-doped indium oxide (ITO) is commonly employed [2], typically about 0.3 μm thick. The glass of chosen thickness may be reinforced to give structural strength. The OTE acts as the conductor of the electrons necessary for the required electron transfer to take place in the electrochromic material. If the latter is the primary electrochrome, a secondary electrochrome may be deposited on the counter electrode. The layers are deposited by electrodeposition, evaporation or some other method directly on to the OTE surface to minimise contact resistances (section 3.5). The use of OTEs for electrochromic applications has been reviewed by Lynam [3] and Granquist [4].

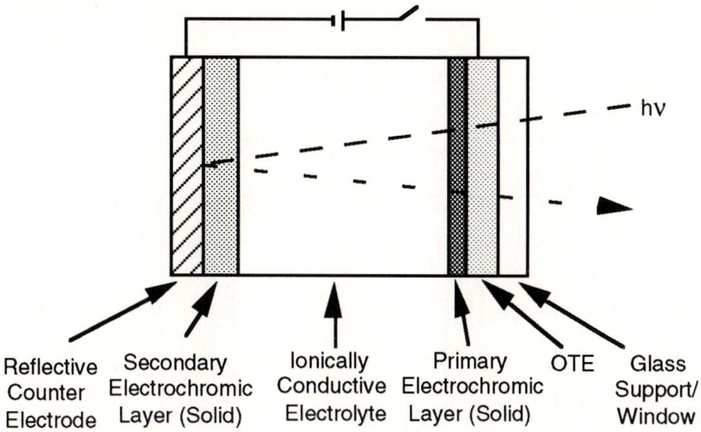

Fig. 3.2 Schematic diagram of an EC mirror operating in reflectance mode (with the reflector as counter electrode).

ECDs operating in a reflective mode (Fig. 3.2) employ a reflective material in the path of the light-beam, both the primary and secondary electrochromes being positioned before it. Examples of reflector include polished platinum [5] or rhodium alloy [6]. The role of reflector and counter electrode are normally combined in this cell with electron transfer proceeding as outlined in chapter 2: the potential is applied between the front OTE and rear electrode/reflector. Displays may have a metallic (e.g. gold) working electrode on which the electrochrome acts, with transparent ITO as counter electrode and window.

Diffusion is usually the predominant mode of mass transport of counter ions in the electrochromic layer while migration is the transport mechanism in the electrolyte.

ECDs in which a separate reflector is placed *before* the secondary layer have been described at length by Baucke [5, 7–9]. Although this arrangement has practical advantages – the counter electrode need not be electrochromic since it is never in the path of the light-beam – the reflector must be ion permeable if charge is to pass across the cell, requiring a porosity which diminishes its reflectivity.

A rear-view car-door mirror operating as Fig 3.2 is illustrated in Fig 1.2.

3.3 All-Solid Cells with Transmissive Operation

Transmissive cells are assembled according to the schematic diagram in Fig. 3.3. The front OTE, the primary electrochromic layer and electrolyte layers operate as outlined in section 3.2 above. Such ECDs are very similar to reflective devices except that the rear electrode obviously cannot be opaque: *all* layers must be fully transparent in the visible spectral range. The apparent intensity change in a transmissive device is only half that of an otherwise identical ECD acting in a reflective mode. This follows since light passes through the primary electrochrome twice in a reflective device, before and after reflection, but only once in a transmissive cell in which the optical path length is thus, in effect, halved.

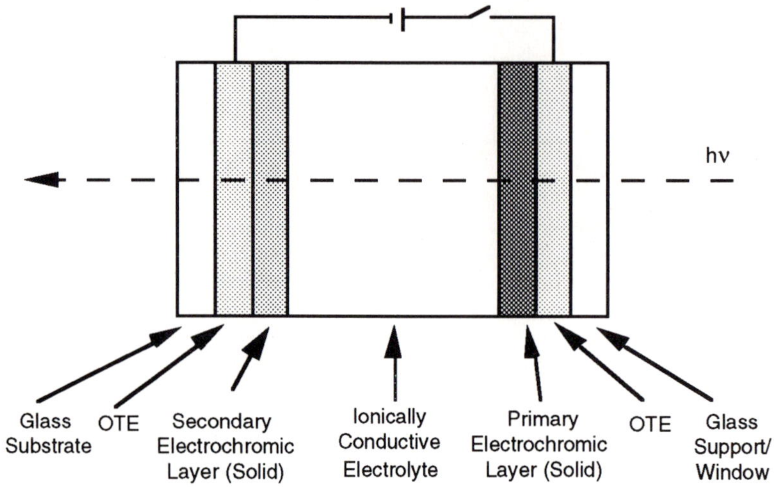

Fig. 3.3 Schematic diagram of an all-solid ECD operating in transmittance mode.

3.4 Solid Electrolytes

Ionically conductive electrolyte, necessarily of negligible electronic conductivity, may be solid, liquid or elastomeric. If liquid electrolytes are used, fluids of high viscosity are usually preferred, for convenience and perhaps safety. The electrolyte layer within a solid (type 3) electrochromic device functions both as an ionic conductor between the electrodes and as a source or sink for ions moving through the electrolyte/electrochrome interfaces during electron transfer. The stability of components to chemical or other (external) attrition determines their utility, thus a choice between solid and liquid systems is required.

Tables 3.1 and 3.2 comprise a selection of solid electrolytes which have been used in ECDs. References [10–12] provide more detail.

Organic electrolytes are often polymers, which fall within two general categories: polymer electrolytes and polyelectrolytes. Firstly, 'polymer electrolytes' are neutral macromolecular species such as poly(ethylene oxide) or poly(propylene glycol) within which salts such as $LiClO_4$, 'triflic acid' CF_3SO_3H or H_3PO_4 are dissolved. Low molecular-weight polymers are liquid, high are almost rigid solids, while those with intermediate molecular weight are highly viscous liquids. These last are used in ECDs and the resultant electrolytes are sometimes termed 'semi-solid solutions' [10].

Table 3.1 Solid or Solid-like Organic Electrolytes for Use in Electrochromic Devices.

ECD System	Electrolyte	Mobile Ion	Ref.
Polymer Electrolytes			
WO_3	CPE + $LiClO_4$	Li^+	[13]
WO_3	Glycerine + $LiClO_4$	Li^+	[14]
WO_3	PEO + $LiClO_4$	Li^+	[15]
V_2O_5	PEO + $LiClO_4$	Li^+	[16]
NiO_x	PEO + $LiClO_4$	Li^+	[17]
WO_3	PEO + H_3PO_4	H^+	[18–20]
WO_3	PVA + H_3PO_4	H^+	[21]
WO_3	PVP + H_3PO_4	H^+	[22]
WO_3	Urea–*p*-TsOH	H^+	[23]
Polyelectrolytes			
IrO_x	Nafion®	H^+	[24, 25]
WO_3	polyAMPS	H^+	[26–29]
WO_3	polyAMPS–PEO	H^+	[30]
WO_3–polyaniline	polyAMPS–polyethylimine	H^+	[31]
WO_3	polyESA	H^+	[25]
WO_3	polyMMA/PEO	Li^+	[32]
	polyMMA/PEO	Li^+	[33, 34]
WO_3	polySSA	H^+	[25, 27]
WO_3	polyVSA	H^+	[28]

Key: CPE = cross-linked polyether; Nafion® = poly(perfluorosulphonic acid); polyAMPS = poly(2-acrylamido-2-methylpropanesulphonic acid); polyESA = poly(ethylene sulphonic acid); polyMMA = poly(methyl methacrylate); polySSA = poly(styrene sulphonic acid); polyVSA = poly(vinyl sulphonic acid); PEO = poly(ethylene oxide); PVA = poly(vinyl alcohol); PVP = poly(vinyl piperidine).

'Polyelectrolytes' are polymers containing ion-labile groups. Examples include polyAMPS: poly(2-acrylamido-2-methylpropanesulphonic acid) which contains proton donor moieties at regular intervals along its backbone.

Table 3.2 Solid Inorganic Electrolytes for Use in Electrochromic Devices.

ECD System	Electrolyte	Mobile Ion	Ref.
a-WO_3	β-Alumina	Na^+	[35, 36]
a-WO_3	Cr_2O_3	H^+	[37]
a-WO_3	Cr_2O_3/V_2O_5	H^+	[38]
WO_3	HUP	H^+	[39]
WO_3	*a*-$LiAlF_4$	Li^+	[40, 41]
a-WO_3	Li_3N	Li^+	[42]
WO_3	$LiNbO_3$	Li^+	[43, 44]
WO_3	MgF_2	H^+	[45–47]
a-WO_3	Nasicon	Na^+	[48]
IrO_x	$PbSnF_4$, PbF_2	F^-	[49]
"WO_3"	PTA	H^+	[50, 51]
WO_3	$RbAgI_5$	Ag^+	[52]
WO_3	Siloxane polymer	Li^+	[53]
WO_3	Ta_2O_5	H^+	[54, 55]
NiO	Water glass	Na^+	[56]
WO_3	ZrO_2	H^+	[54]

Key: HUP = hydrogen uranyl phosphate; nasicon = sodium silicon conductor: $Na_{(1+x)}Zr_2Si_xP_{(3-x)}O_{12}$; PTA = phosphotungstic acid; water glass = semi-solid sodium silicate solution.

Semi-solid organic electrolytes (table 3.1) are much in vogue owing to the relative ease of device fabrication. ECDs containing such polymer electrolytes may be more conveniently constructed than are solid systems, since primary and secondary electrochromes are prepared separately on their respective conducting substrates. The two halves of the device are then joined using intervening electrolyte as an adhesive layer between them. Such electrolyte layers are not brittle and, furthermore, the stresses caused by the effects of expansion *etc.* of solid electrochromes is readily borne since such electrolytes are elastomeric (see below). These materials do, however, have a slight proclivity to photolytic degradation, and the strict control of internal moisture levels in these materials is often essential.

Many solid oxides serve as media for cation conduction, which is greatest for H^+ and often appreciable for Li^+. Solid inorganic electrolytes (table 3.2) are superior to organic systems in their enhanced stability to photolytic degradation. However, in ECDs with solid electrolytes, significant interfacial contact resistances are difficult to avoid. In a recent WO_3-based prototype, for example [8], electrolyte was evaporated onto the secondary electrochrome with the primary electrochrome deposited as a further layer. ITO as electronic conductor was then sputtered onto this sandwich as yet another layer. Production costs were too high to allow manufacture, however.

A second deficiency of solid electrolytes is the relative fragility of these often brittle solids. Since the electrochromes, both primary and secondary, generally expand or contract during ion insertion or egress, strains are incurred which cause eventual disintegration of the oxide electrolyte sandwiched between. Thus, the bulk densities of cobalt oxide are 5.18 g cm^{-3} for CoO and 6.45 g cm^{-3} for the oxidation product, Co_2O_3 [57]. There is no reason to suppose that the densities of the respective thin films differ significantly from the bulk values.

3.5 The Preparation of Solid Electrochromic Films

Many species are employed as permanently insoluble films prepared by electrodeposition. Examples include the oxides of cobalt [58, 59], iridium [60–63], molybdenum [64], nickel [56, 65–69] or tungsten [70–73], all formed from aqueous solution. Prussian blue [74, 75] may also be prepared this way, although special conditions are necessary to effect electrodeposition. A technique relying on the dissolution of a sacrificial anode has also been used to deposit films of Prussian blue [76, 77].

The standard text detailing the techniques for vacuum deposition of electrochromic materials in thin film form remains 'Vacuum Deposition of Thin Films' by Holland, 1956 [78]. Thermal evaporation is usually employed to prepare films of metal, metal oxide or phthalocyanine, organic species may also be be evaporated [79]. The procedure is as follows. Bulk material is electrically heated while in a foil boat* within the vacuum chamber, and sublimes onto the substrate (usually ITO-coated) glass. The chamber is commonly evacuated to 10^{-6} mmHg, while the evaporation is performed at 10^{-5} mmHg.

* Evaporation of the metal of the boat can occur when a tungsten or molybdenum foil is used as the boat material [78]. Tungsten boats can add as much as 5 % W to the film being evaporated and the more involatile Mo may also sublime and be included within WO_3 [86].

The substrate may be placed on a cooled holder to minimise the annealing effects of heating. (Annealing causes crystallisation of amorphous films, which may be undesirable since diffusion through crystalline materials can be relatively slow – see section 4.7.2.)

When the sputtering method (RF or DC-magnetron) is used to prepare films of metal oxide, a target of the metal (e.g. W, Ni or W, of *ca.* 10 cm diameter) is bombarded with reactive oxygen atoms from an O_2/N_2 or O_2/Ar gas mixture at low pressures, for example, 7×10^{-5} to 7×10^{-2} mmHg (10^{-2} to 10 Pa). Since sputtering usually involves a power of 100 W or so [80], substrates are supported on a water-cooled mount to prevent thermal annealing and possible thermal evaporation of the growing thin film.

When the electron-beam sputtering (EBS) technique is used, the sample is placed on an inductively heated cermet (ceramic metal), and ionised with a tungsten filament (typically at 5 kV to give a sputtering current of 1.5 mA). Such polycrystalline films are usually sub-stoichiometric, for example, the 'MoO_3' prepared by electron-beam sputtering is in fact $MoO_{(3-y)}$ where y is *ca.* 0.3.

Chemical vapour deposition (CVD) can be used to prepare thin films of WO_3 and MoO_3, and is somewhat similar to conventional thermal evaporation. Following sublimation of the starting material, $W(CO)_6$ or $Mo(CO)_6$ [81] (or vaporisation, of volatile liquid), the vaporised samples react with (i.e. are oxidised by) trace amounts of oxygen gas bled deliberately into the chamber. The immediate product of such CVD is black, consisting chiefly of finely divided metal. Carbon monoxide waste, together with excess O_2, is removed rapidly by the vacuum pump. Thermal oxidation in air at *ca.* 500–600 °C is then required. For this reason, metal oxide thin films formed via CVD are generally polycrystalline.

The sol-gel process [82] is particularly suited to the production of large area ECDs, for applications such as smart glass windows [83]. Here, gelatinous material is applied to the conducting working electrode, the gel being formed via sol resulting from partial precipitation. Gel is applied on to the substrate prior to thermal curing, which drives off solvent and/or chemically modifies the component(s) within the gel.

Livage *et al.* [83–85] prepared WO_3 films from a sol-gel intermediate, using tungstic acid in aqueous solution as precursor: ethanol was added to aqueous H_2WO_4 to form colloid of mean particle diameter 0.5–1.0 μm which was converted to gel. Final production of thin-film WO_3 was achieved by spraying the gel on to hot ITO-coated glass. A recent review by Agrawal, Cronin and Zhang [87] details the solid-state electrochromic films which may be prepared with the sol-gel process.

The dip-coating (or 'spin coating') procedure is simpler still: substrate is dipped into solutions containing either colloidal or solvated species. Emersion, and thermal drying, forms the desired product. Multiple layers may be prepared in this way. An example is WO_3 or MoO_3 [88, 89] made by dipping electrodes in a solution itself prepared using hydrogen peroxide to dissolve tungsten or molybdenum metal respectively. Such electrochromic films on Pt or ITO glass are reportedly of good quality [88]. Spin coating is similar to dip coating except the solvent is removed by rapid spinning of the sample rather than simple thermal drying. Spin-coating tends to yield films with more even thicknesses.

3.6 Liquid Electrolytes

For liquid-containing ECDs – types 1 and 2 – as with solid-electrolyte cells, ionic charge carriers are required in the inter-electrode region. Any suitable unreactive salt will serve: it must ionise sufficiently in the chosen solvent, it should not precipitate the electrochrome if this is not wanted, and it must be stable especially against photolysis. Often DMF, and PC are used as solvent, which clearly must dissolve the electrochromes and any added salt. $LiClO_4$ is commonly used as the salt, though perchlorates in organic media need careful handling especially against explosion on drying out; $LiBF_4$ is a useful alternative. The choice of electrolyte composition is largely dictated by the nature of the electrochrome. The preparation of electrolytes is detailed by Sawyer and Roberts [90].

In the operation of special devices, like the electrochromic rear-view mirror described in the next section, migration (see section 2.5.3.1) of the ionic solution-phase electrochromes in the field applied across the electrodes could be an intrinsic part of the ECD mechanism and, here, the addition of extra inert electrolyte is obviously to be avoided.

3.7 Self-darkening Electrochromic Rearview Mirror for Cars Employing Type 1 (Solution-phase) Electrochromes

While a number of designs of an automatically self-darkening car rearview mirror, to avoid dazzle from the headlights of a following car, have been proposed and put into effect, probably the greatest production has been that by Gentex, a company which has developed a particularly simple construction. Basically, an ITO-glass surface (conductive side inwards) and the reflective metallic surface, spaced a fraction of a millimetre apart, form the two electrodes of the cell, with a solution containing two electrochromes in between. The details of the compositions are buried in turgid patents [91–94], selection from which, to envisage the precise systems actually in use, is virtually impossible. However, it is possible to infer the operation as follows. A substituted bipyridilium species, undoubtedly cationic ('B^+'), serves as cathodic electrochrome. When the mirror is switched on, the positive charge that the uncoloured initial species bears will thus under ohmic migration drive it to the cathode, to undergo reductive colouration. The other electrochrome (to be oxidised) may be a substituted radical or molecular thiazine 'TA^-', or a phenylene diamine, initially bearing a negative charge, for the propulsion by the electric field to the anode. The electrochromes thus appears to act also as electrolyte, and would require a solvent of high permittivity to minimise ion pairing of B^+ and TA^-. Oxidation of TA^- at the anode then evokes the methylene-blue type of colouration. In operation, the colour in a commercially available mirror is indeed a very deep green-blue.

While not an electrochromic phenomenon, the ingenious control system is worth noting. A photosensitive detector is placed facing rearward to monitor any dazzling incident light but, in daylight, it would be triggered, resulting in an unwanted darkening of the mirror. This is avoided by the second *forward-looking* detector, which on seeing daylight, is programmed to cancel any operation of the controlling sensor, which thus responds only in the dark of night.

To return to the electrochemistry, we have envisaged the reduction $B^+ + e^- \rightarrow B^o$ and complementary oxidation $TA^- \rightarrow TA^o + e^-$ in a dual electrochromic colouration process. The products will thus diffuse (and suffer convection) away from their respective electrodes and meet in the intervening solution. Here a mutual reaction regenerating the original uncoloured species must ensue:

$$B^o + TA^o \rightarrow B^+ + TA^- \qquad (3.1)$$

Thus maintenance of colouration requires application of a continuous current which is, however, very small, to replenish the coloured electrochromes lost by their mutual redox reaction in solution. This aspect of the reaction represents a slight divergence from one of the claimed benefits of electrochromism, that switching the colour on or off are the only current-consuming operations, but the outcome here is a strictly practical device of considerable robustness. Since 1990 [95], great numbers are claimed to have been in use, and one must await the test of time – durability over perhaps a decade is clearly desirable.

References

[1] T.P. Brody and P.R. Malmsberg, *J. Hybrid Microelec.*, II (1979) 29.

[2] R.W. Murray, W.R. Heineman and C.W. O'Dom, *Anal. Chem.*, 39 (1967) 1666.

[3] N.R. Lynam, *Proc. Electrochem. Soc.*, 90–2 (1990) 201.

[4] C.G. Granqvist, *Appl. Phys. A*, A57 (1993) 19.

[5] F.G.K. Baucke, *Rivista della Staz. Sper. Vetro,* 6 (1986) 119.

[6] W. Wagner, F. Rauch, C. Ottermann and K. Bange, *Nuc. Instr. Meth. Phys. Res.*, B50 (1990) 27.

[7] F.G.K. Baucke, K. Bange and T. Bange, *Displays*, October (1988) 179.

[8] F.G.K. Baucke, *Proc. Electrochem. Soc.*, 90–2 (1990) 298.

[9] F.G.K. Baucke, *S.P.I.E. Institute Series*, IS4 (1990) 518.

[10] P. Hagenmuller and W. van Gool (eds.), 'Solid Electrolytes: General Principles, Characterisation, Materials, Applications', Academic Press, London, 1978.

[11] D. Deroo, in B. Scrosati (ed.), 'Second International Symposium on Polymer Electrolytes', Elsevier Applied Science, London, 1990. p 433.

[12] B. Scrosati, in B. Scrosati (ed.), 'Applications of Electroactive Polymers', Chapman and Hall, London, 1993. Chapter 8.

[13] F. Croce, S. Passerini, A. Selvaggi and B. Scrosati, *Solid State Ionics,* 40–41 (1990) 375.

[14] F. Kanai, S. Kurita, S. Sugioka, M. Li and Y. Mita, *J. Electrochem. Soc.*, 129 (1982) 2633.

[15] S. Passerini, B. Scrosati, A. Gorenstein, A.M. Andersson and C.G. Granqvist, *J. Electrochem. Soc.*, 136 (1989) 3394.

[16] A.M. Andersson, C.G. Granqvist and J.R. Stevens, *Proc. S.P.I.E.*, 1016 (1988) 41.

[17] P. Baudry and D. Deroo, *Proc. Electrochem. Soc.,* 90–2 (1990) 274.

[18] B. Scrosati, in B.V.R. Chowdari and S. Radhakrishna (eds.), 'Proceedings of the International Seminar on Solid State Ionic Devices', World Scientific Publ. Co., Singapore, 1988. p 341.

[19] P. Pedone, M. Armand and D. Deroo, *Solid State Ionics*, 28–30 (1988) 1729.

[20] J.A. Duffy, M.D. Ingram and P.M.S. Monk, *Solid State Ionics*, 58 (1992) 109.

[21] R.P. Singh, S.L. Agrawal and V.P. Singh, *Extended Abstracts of the 7th International Conference on Solid State Ionics*, 1989, abstract, 6pB–35.

[22] M. Armand, D. Deroo and D. Pedone in ref [21] p 515.

[23] M. Shizukuishi, *Jpn. J. Appl. Phys.*, 20 (1981) 581.

[24] G. Beni and C.E. Rice, in P. Vashishta, J.N. Mundy and G.K. Shenoy (eds.), 'Fast Ion Transport in Solids: Electrodes and Electrolytes', Elsevier, Amsterdam, 1979. pp. 75 and 99.

[25] J.-P. Randin, *J. Electrochem. Soc.*, 129 (1982) 1215.

[26] C.G. Granqvist and J.S.E.M. Svensson, *Solar Energy Mater.,* 16 (1987) 19.

[27] R.D. Giglia and G. Haake, *Proc S.I.D.*, 23 (1982) 41.

[28] J.-P. Randin, *Electronics*, 54 (1981) 89.

[29] G. Beni, *Solid State Ionics,* 4 (1981) 157.

[30] S.F. Cogan and R.D. Rauh, *Solid State Ionics*, 28–30 (1988) 1717.

[31] M. Akhtar and H.A. Weakliem, *Proc. Electrochem. Soc.*, 90–2 (1990) 232.

[32] M. Nishikawa, H. Ohno, N. Kobayashi, E. Tsuchida and R. Hirohashi, *J. Soc. Photog. Sci. Technol. Jpn.*, 51 (1988) 184.

[33] M. Nishikawa, H. Ohno, N. Kobayashi, E. Tsuchida and R. Hirohashi, *J. Soc. Photog. Sci. Technol. Jpn.*, 51 (1988) 375.

[34] N. Kobayashi, R. Hirohashi, H. Ohno and E. Tsuchida, *Solid State Ionics,* 40–41 (1990) 491.

[35] M. Green and K.S. Kang, *Thin Solid Films,* 40 (1977) 49.

[36] M. Green and K.S. Kang, *Solid State Ionics,* 4 (1981) 141.

[37] E. Inoue, K. Kawaziri and A. Izawa, *Jpn. J. Appl. Phys.*, 16 (1977) 2065.

[38] M. Shizukuishi, I. Shimizu and E. Inoue, *Jpn. J. Appl. Phys.*, 20 (1981) 575.

[39] A.T. Howe, S.H. Sheffield, P.E. Childs and M.G. Shilton, *Thin Solid Films,* 67 (1980) 365.

[40] S.F. Cogan, T.D. Plante, E.J. Anderson and R.D. Rauh, *Proc. S.P.I.E.*, 562 (1985) 23.

[41] T. Oi, *Ann. Rev. Mater. Sci.*, 16 (1986) 185.

[42] M. Miyamura, S. Tomura, A. Imai and S. Inomata, *Solid State Ionics*, 4 (1981) 149.

[43] R.B. Goldner, T.E. Haas, K.K. Wong, P. Norton, G. Foley, G. Berera, G. Wei, S. Schutz and R. Chapman, *Solid State Ionics*, 28–30 (1988) 1715.

[44] R.B. Goldner, G. Seward, K. Wong, T. Haas, G.H. Foley, R. Chapman and S. Schutz, *Solar Energy Mater.*, 19 (1989) 17.

[45] A. Deneuville, P. Gérard and R. Billat, *Thin Solid Films*, 70 (1980) 203.

[46] P. Gérard, A. Deneuville and R. Courths, *Thin Solid Films*, 71 (1980) 221.

[47] T. Yoshimura, M. Watanabe, K. Kiyote and M. Tanaka, *Jpn. J. Appl. Phys.*, 21 (1982) 128.

[48] G.G. Barna, *J. Electron. Mater.*, 8 (1979) 153.

[49] C.E. Rice and P. Bridenbaugh, *Appl. Phys. Lett.*, 38 (1981) 59.

[50] B. Tell and F. Wudl, *J. Appl. Phys.*, 50 (1979) 5944.

[51] B. Tell, *J. Electrochem. Soc.*, 127 (1980) 2451.

[52] M. Green and D. Richman, *Thin Solid Films*, 24 (1974) S45.

[53] J.R. Stevens, J.S.E.M. Svensson, C.G. Granqvist and R. Spindler, *Appl. Opt.*, 26 (1987) 3489.

[54] Y. Hajimoto, M. Matsushima and S. Ogura, *J. Electron. Mater.*, 8 (1979) 301.

[55] T. Saito, Y. Ushio, M. Yamada and T. Niwa, *Extended Abstracts of the 7th International Conference on Solid State Ionics*, 1989, abstract 6pB–40.

[56] C. Liquan, D. Ming, C. Yunfa, S. Chunxiang and X. Rungjian, *Extended Abstracts of the 7th International Conference on Solid State Ionics, Japan*, 1989, abstract 6pB–38.

[57] R.C. Weast (ed.), 'Handbook of Chemistry and Physics', 66th Edn., CRC Press, Boca Raton, Florida, 1986. page B–90.

[58] A. Gorenstein, C.N. Polo da Fonsecu and R. Torresi, *Proc. S.P.I.E.*, 1536 (1991) 104.

[59] P.M.S. Monk, D.S. Higham and S.L. Chester, *Proc. Electrochem. Soc.*, 94–2 (1994) 100.

[60] S. Gottesfeld, J.D.E. McIntyre, G. Beni and J.L. Shay, *Appl. Phys. Lett.*, 33 (1978) 208.

[61] S. Gottesfeld and J.D.E. McIntyre, *J. Electrochem. Soc.*, 126 (1979) 742.

[62] G. Beni, C.E. Rice and J.L. Shay, *J. Electrochem. Soc.*, 127 (1980) 1342.

[63] K.S. Kang and J.L. Shay, *J. Electrochem. Soc.*, 130 (1983) 766.

[64] A. Guerfi and L.H. Dao, *J. Electrochem. Soc.*, 136 (1989) 2435.

[65] W. Visscher and E. Barendrecht, *J. Electroanal. Chem.*, 154 (1983) 69.

[66] M.K. Carpenter, R.S. Connell and D.A. Corrigan, *Solar Energy Mater.*, 16 (1987) 433.

[67] R.M. Bendert and D.A. Corrigan, *J. Electrochem. Soc.*, 136 (1989) 723.

[68] R.M. Bendert and D.A. Corrigan, *J. Electrochem. Soc.*, 136 (1989) 1369.
[69] P.-C. Yu, G. Nazri and C.M. Lampert, *Solar Energy Mater.*, 16 (1987) 1.
[70] M.T. Nguyen and L.H. Dao, *Proc. Electrochem. Soc.*, 90–2 (1990) 246.
[71] P.K. Shen and A.C.C. Tseung, *J. Mater. Chem.*, 2 (1992) 1141.
[72] P.K. Shen, J. Syed-Bokhari and A.C.C. Tseung, *J. Electrochem. Soc.*, 138 (1991) 1778.
[73] P.M.S. Monk and S.L. Chester, *Electrochim. Acta*, 38 (1993) 1521.
[74] V.D. Neff, *J. Electrochem. Soc.*, 125 (1978) 886.
[75] R.J. Mortimer and D.R. Rosseinsky, *J. Electroanal. Chem.*, 151 (1983) 133.
[76] K.-C. Ho, *Proc. Electrochem. Soc.*, 94–2 (1994) 170.
[77] K.-C. Ho, T.G. Rukavina and C.B. Greenberg, *Proc. Electrochem. Soc.*, 94–2 (1994) 252.
[78] L. Holland, 'Vacuum Deposition of Thin Films', Chapman and Hall, London, 1956.
[79] A. Yasuda and J. Seto, *J. Electroanal. Chem.*, 283 (1990) 197.
[80] C.G. Granqvist and J.S.E.M. Svensson, *Solar Energy Mater.*, 16 (1987) 19.
[81] A. Donnadieu, D. Davazoglou and A. Abdellaoui, *Thin Solid Films*, 164 (1988) 333.
[82] L.L. Hensch and J.K. West, *Chem. Rev.*, 90 (1990) 33.
[83] J. Livage, A. Zarudiansky, R. Rose and P. Judenstein, *Solid State Ionics*, 28–30 (1988) 1722.
[84] A. Chemseddine, R. Morineau and J. Livage, *Solid State Ionics*, 9–10 (1983) 357.
[85] A. Chemseddine, M. Henry and J. Livage, *Rev. Chemie. Minér*, 21 (1984) 487.
[86] M.R. Goulding and C.B. Thomas, *Thin Solid Films*, 62 (1979) 175.
[87] A. Agrawal, J.P. Cronin and R. Zhang, *Solar Energy Mater.*, 31 (1993) 9.
[88] K. Hinokuma, K. Ogasawara, A. Kishimoto, S. Takana and T. Kudo, *Solid State Ionics*, 53–56 (1992) 507.
[89] S. Takano, A. Kishimoto, K. Hinokuma and T. Kodo, *Solid State Ionics*, 70/71 (1994) 636.
[90] D.T. Sawyer and J.L. Roberts, 'Experimental Electrochemistry for Chemists', Wiley, New York, 1974.
[91] F.T. Bauer and J.H. Bechtel, Gentex Corp., U.S. Patent, 4,443,057 (1984).
[92] H.J. Byker, Gentex Corp., U.S. Patent, 4,902,108 (1990).
[93] J.H. Bechtel and H.J. Byker, Gentex Corp., U.S. Patent, 4,917,477 (1990).
[94] H.J. Byker, Gentex Corp., U.S. Patent, 5,128,799 (1992).
[95] H.J. Byker, *Proc. Electrochem. Soc.*, 94–2 (1994) 3.

Part II

Electrochromic Systems

General Introduction

In part II, we discuss electrochromic systems which have been considered for inclusion in ECDs. There are many systems within this corpus which have received great interest from the start, but also many on which still only little work has been published.

The most recent reviews of electrochromes and their use in electrochromic devices are 'Electrochromic Devices' by Mastragostino [1], 'Electrochromic Materials: Microstructure, Electronic Bands and Optical Properties' by Granqvist [2] and 'Chemogenic Materials: Electrochromism' by Bohnke [3], all three appearing in 1993. Granqvist also published two short reviews in 1990 [4, 5] and two works [6, 7] by Scrosati (shorter than [1]) appeared in 1992 and 1990. A work of limited scope by Oi [8] appeared in 1986 and Lampert's [9] more comprehensive study appeared in 1984, but is aimed chiefly at the construction of smart windows. Chang, Gilbert and Sun [10] produced a brief study in 1975. Scrosati's 'Laminated Electrochromic Displays and Windows' [11] appeared in 1993. Reviews by Lusis [12] (in Russian, 1981), Matsuhiro and Matsuda [13] (in Japanese, 1984), and Bludska and Vondrak [14] (in Czech, 1991) are also available.

Dautremont-Smith published two excellent, comprehensive reviews of practical electrochromic systems, in 1982: 'Oxides with Cathodic Colouration' [15] and 'Oxides with Anodic Colouration' [16]. 'Electrochromic Displays based on WO_3' by Faughnan and Crandall [17] (1980) remains the best discussion of the criteria for ECD operation (not just for WO_3), and is both readable and still largely relevant. 'A Review on Electrochromic Devices for Automotive Glazing' by Demiryont appeared in 1991 [18].

Finally, two general surveys of electrolytes are available: 'Electrochromic Windows and Displays using Polymer Electrolytes' [19] by Deroo appeared in 1990, and 'Applications of Proton Conductors in Electrochromic Devices (ECDs)' by Bohnke in 1992 [20].

Potentials cited here have been converted to the saturated calomel electrode (SCE) potential scale, when aqueous electrolyte was used. This attempt at standardisation unfortunately will have involved reversing the procedures followed by some authors, of *citing* potentials with respect to zero for a 'standard hydrogen electrode', for values *measured* with respect to an SCE, then 'corrected' to the hydrogen scale. We have used the value of 0.242 V for the SCE on the hydrogen scale [21].

References

[1] M. Mastragostino, in B. Scrosati (ed.), 'Applications of Electroactive Polymers', Chapman and Hall, London, 1993. Chapter 7.

[2] C.G. Granqvist, *Appl. Phys.*, A56 (1993) 1.

[3] O. Bohnke, in 'Kirk-Othmer Encyclopedia of Chemical Technology', 4th edn., vol. 6, John Wiley and Son, Chichester, 1993, p 312.

[4] C.G. Granqvist, *Crit. Rev. Solid State Mater. Sci.*, 16 (1990) 291.

[5] C.G. Granqvist, *Solid State Ionics*, 53–56 (1992) 479.

[6] B. Scrosati, in B.V.R. Chowdari, (ed.), 'Solid State Ionics: Materials and Applications', World Publishing Co. Ltd., Singapore, 1992. p 321.

[7] B. Scrosati, *Mol. Cryst. Liq. Cryst.*, 190 (1990) 161.

[8] T. Oi, *Ann. Rev. Mater. Sci.*, 16 (1986) 185.

[9] C.M. Lampert, *Solar Energy Mater.*, 11 (1984) 1.

[10] I.F. Chang, B.L. Gilbert and T.I. Sun, *J. Electrochem. Soc.*, 122 (1975) 955.

[11] B. Scrosati in ref. [1], chapter 8.

[12] A. Lusis, *Oksid Electrokhrom Materialy Riga,* (1981) 13, cited in *Chem. Abstr.* 97: R153,801*q*.

[13] M. Matsuhiro and Y. Masuda, *Kagaku. Zohan (Kyoto),* 104 (1984) 107, cited in *Chem. Abstr.* 103: R96,116*m*.

[14] J. Bludska and J. Vondrak, *Chem. Listy,* 85 (1991) 776, cited in *Chem. Abstr.* 115: 265221*x*.

[15] W.C. Dautremont-Smith, *Displays*, 3 (1982) 3.

[16] W.C. Dautremont-Smith, *Displays,* 3 (1982) 67.

[17] B.W. Faughnan and R.S. Crandall, in J.I. Pankove (ed.), 'Display Devices', Springer-Verlag, Berlin, 1980, chapter 5.

[18] H. Demiryont, *Proc S.P.I.E.*, 1536 (1991) 2

[19] D. Deroo, in B. Scrosati (ed.), 'Proceedings of the Second International Symposium on Polymer Electrolytes', Elsevier, London, 1990. p 433.

[20] O. Bohnke, in P. Columban (ed.), 'Proton Conductors: Solids, Membranes and Gels – Materials and Devices', Cambridge University Press, 1992. Chapter 38.

[21] D.I. Hitchman and A.C. Taylor, *J. Am. Chem. Soc.*, 59 (1937) 1813.

A Inorganic Systems

4 Metal Oxides

4.1 Introduction – Colour in Mixed-valence Systems

The intense optical absorption shown by the metal oxides in this chapter is due to optical *intervalence* charge transfer (CT) [1–4]. Here, an electron is excited to a similar, vacant, orbital on an adjacent ion or molecule (Section 1.4.2). Consider tungsten trioxide, WO_3: before electroreduction, the oxide is a pale yellow in hue, all tungsten sites having identical oxidation states of +VI. Electron transfer from an electrode to W^{VI} (reduction) forms some W^{V} centres, and the desired blue form of the electrochrome is produced. Partial reduction introducing negative charge requires charge compensation by cation insertion (say of H^+), with a fractional insertion coefficient x, to form $H_xW^{V}{}_xW^{VI}{}_{(1-x)}O_3$. Hence, the W^V centres in the presence of W^{VI} allow a colour-generating photo-effected intervalence transition to occur between the different tungsten sites

$$\overset{\mathrm{V}}{\mathrm{W}}_{\mathrm{A}} + \overset{\mathrm{VI}}{\mathrm{W}}_{\mathrm{B}} \xrightarrow{h\nu} \overset{\mathrm{VI}}{\mathrm{W}}_{\mathrm{A}} + \overset{\mathrm{V}}{\mathrm{W}}_{\mathrm{B}} \tag{4.1}$$

Such intervalence transitions are characterised by broad, intense and relatively featureless absorption bands in the UV, visible or near IR regions, with large molar absorptivities, for example, ε for H_xWO_3 lies in the range 1,400–5,600 $dm^3\ mol^{-1}\ cm^{-1}$, the value being dependent on x [5]. Intervalence transitions appear only when both redox states are present, so *complete* reduction of the W^{VI} sites to form W^V would not result in more intense absorption since only one of the two valence species would be present. In any case, a new phenomenon intervenes: when the interaction between the sites is strong enough to allow delocalisation of the transferable electrons via the formation of metal-like conduction bands, metallic properties such as reflection replace optical absorption, as occurs here when $x > ca.\ 0.3$.

The broadness of many optical intervalence charge-transfer bands means that metal oxide ECDs are almost invariably blue-grey or black, since any reflected colour is hence of feeble intensity; the mixed-valence oxide of cobalt is brown [6], however.

There is a substantial literature [1–4] on mixed-valence systems in general, mostly nonmetallic, many of which could be made electrochromic. Homonuclear systems (comprising different oxidation states of the same element) and heteronuclear systems (comprising different elements in suitable oxidation states to permit optical CT) are encompassed. The question of *initial* delocalisation of the transferable electron (i.e. prior to any optical CT) provides a focus of much of the study. Robin and Day have used the extent of such delocalisation (minimal, moderate or substantial) to delineate categories of intervalence (Groups I, II or III respectively). Group III – substantial delocalisation – falls into either IIIA, delocalisation within clusters, each virtually isolated from the rest, or IIIB, substantial extensive delocalisation. Group IIIB shows metallic properties, as exemplified by the $W^{VI/V}$ system referred to above.

The metals in this chapter follow alphabetically, except that section 4.9 covers systems too little studied to merit a separate discussion for each. Finally, the new and growing field of mixed metal oxides (section 4.10) exemplifies the advantages of using more than one metal in the oxide electrochrome.

All electrode reactions have been cited as proceeding in the observed direction of colouration; in electrochemistry texts, electrode reactions are by convention all written as electron acquisitions (reduction) (chap. 2).

4.2 Cobalt Oxide

Films of cobalt oxide, grown anodically on cobalt metal in 1 M NaOH electrolyte, are blue [7], the colour soon becoming brown on standing [8]. Burke and Murphy [7, 8] quote the reaction as involving

$$\underset{\text{(blue)}}{3Co_2O_3} + H_2O + 2e^- \rightarrow \underset{\text{(brown)}}{2Co_3O_4} + 2\,OH^- \qquad (4.2)$$

Behl and Toni [9] state that many electrochromic colours arise presumably from varying oxide/hydroxide compositions and composition-dependent CT absorptions including white, pink, brown and black, in agreement with Benson [10]. Below 1.47 V, films are orange (or yellow-brown) but above this potential the films become dark brown, (or black for thicker films). The orange form of the oxide may also contain hydrated $Co(OH)_2$, probably formed after proton uptake; on Co metal anodised in 0.1 or 1.0 M NaOH, this oxide is the predominantly low-valence product, detected by FTIR measurement [11]. The identity of the Co^{III} oxide(s) formed by oxidation of $Co(OH)_2$ could not be assigned conclusively by FTIR [11]. The authors also quote IR data for all the oxides of cobalt (including those above together with CoO and CoO.OH).

Alternatively, films of $LiCoO_2$ may be prepared by thermal evaporation, RF-sputtering, spray pyrolysis, or sol-gel techniques [12]. The product of RF-sputtering, using powdered $LiCoO_2$ and slow deposition, has the best optical properties [13]. Such films are polycrystalline [14, 15]. Electron transfer to the $LiCoO_2$ electrochrome, immersed in a $LiClO_4$–propylene carbonate solution, results in an electrochromic colour change from deep brown to transparent, accompanied by removal of Li^+ [16].

Electrochromic films of cobalt oxide may also be formed by electrodeposition from aqueous solution [17–19] using simple solutes such as $Co(NO_3)_2$ or $CoCl_2$ in the initial deposition solution (e.g. using a 1.0 M precursor in water). A novel green product (oxygen rich CoO) is reportedly attainable by this method – one of the few wholly inorganic systems to show such a colour – and is formed using negative potentials [19]. For such electrodeposited material, the electrochromic transition is green-to-brown:

$$\underset{\text{(green)}}{3CoO} + 2OH^- \rightarrow \underset{\text{(brown)}}{Co_3O_4} + 2e^- + H_2O \tag{4.3}$$

In a different electrodeposition procedure cobalt is oxidatively dissolved in H_2O_2 [19]. The electrochromism of cobalt(III) oxyhydroxide CoO•OH has also been described [17].

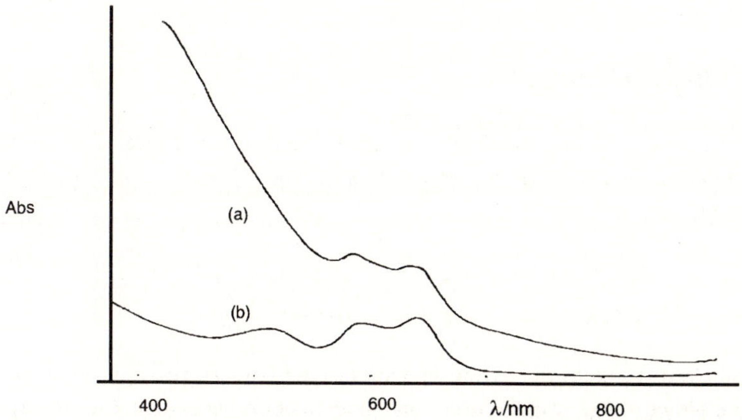

Fig. 4.1 *UV-vis spectra of electrodeposited cobalt oxide on ITO: (a) dark brown Co_3O_4 and (b) green CoO. (Figure reproduced from [19] by permission of The Electrochemical Society).*

4.3 Indium Tin Oxide

Indium(III) oxide, when doped with *ca.* 8 % w/w tin (ITO or 'indium tin oxide') is a familiar semiconducting thin film used as a conductive coating in the construction of optically transparent electrodes [20–23]. The intrinsic electrochromism of ITO is weak, the ITO film being colourless when oxidised and a very pale brown colour following reduction. Steele *et al.* [24–26] quote the *CR* of ITO as 7:1 with Li^+ as the inserted ion.

$$\underset{\text{(colourless)}}{In_2O_3} + 2x\,(Li^+ + e^-) \rightarrow \underset{\text{(pale brown)}}{Li_{2x}\overset{III}{In}_{(1-x)}\overset{I}{In}_xO_3} \tag{4.4}$$

Irreversible insertion can cause the failure of ITO-based devices. The lack of reversibility may possibly be due to the formation of a thin surface layer of In^{I} as reduction product on the electrolyte-facing side of the film [27].

Cogan *et al.* [28] and Goldner and Yu *et al.* [29–31] have also studied the insertion of Li^+ into ITO, while Svensson and Granqvist [32] have examined ITO films prepared by electron beam sputtering; the usual method of preparation is to use RF-sputtering [33, 34].

The thickness of the ITO layer and the attendant electron delocalisation (adjusted by the preparation procedure and the Sn:In ratio) may be optimised to adjust the transmittance of an ITO sheet between 10–78 % [35].

ITO has been incorporated into ECDs as a counter electrode material [24, 25, 36–38] rather than as the primary electrochrome, because of its inherently weak electrochromism. Also, the movement of Li^+ species through the lattice is rather slow ($D = 6 \times 10^{-13}$ to 5×10^{-14} $cm^2\ s^{-1}$ [24], or 1×10^{-11} $cm^2\ s^{-1}$ [30]), which may also be the cause of hysteresis in coulometric titration curves [30].

Further operational problems also occur. Moisture levels must be minimised if ITO is to be used as a counter electrode since with H_2O present, over-reduction occurs readily, forming elemental metal as the product. The use of anhydrous electrolyte overcomes this problem. Secondly, low-charge metal species (perhaps as oxo-anions) within the reduced ITO are relatively soluble and leach into solution [39–41], particularly in the presence of aqueous acids.

4.4 Iridium Oxide

Most of the published work on films of iridium oxide was performed by Beni, Gottesfeld, McIntyre, Rice and Shay of Bell Laboratories. The iridium oxide used in ECD devices is actually hydrated, as iridium hydroxide, $Ir(OH)_3$. The mechanism of colouration is still uncertain, and two different reaction routes have been proposed. The first is based on cation loss [42, 43]

$$\underset{\text{(colourless)}}{Ir(OH)_3} \rightarrow \underset{\text{(blue/grey)}}{IrO_2{\bullet}H_2O + H^+ + e^-} \tag{4.5}$$

while the second is based on anion insertion [41]

$$Ir(OH)_3 + OH^- \rightarrow IrO_2{\bullet}H_2O + H_2O + e^- \tag{4.6}$$

Irrespective of whether the mechanism is hydroxyl insertion or proton extraction, $Ir(OH)_3$ is the bleached form of the oxide and the coloured form is IrO_2 [44–46].

Although sputtering of iridium on to an OTE in an oxygen atmosphere has been used [47] to make iridium oxide films, there are only two methods of film preparation commonly used in current practice: electrochemical growth, making anodic iridium oxide films ('AIROFs') and sputtered iridium oxide films ('SIROFs').

Anodically grown films [42–45, 47–51] are made by the potentiostatic cycling between −0.25 V and +1.25 V of an electrode in an iridium-containing solution, commonly in an electrolyte of 0.5 M H_2SO_4, [42].

AIROFs made by this route have a contrast ratio reportedly as high as 70:1 and quoted response times of 20–40 ms, which are faster than for tungsten or vanadium oxides in ECDs. Such AIROFs may be cycled for more than 10^6 times. Solid state AIROFs have been made in which the electrolyte is polymeric, but these have slower response times [51]. For the coloured blue-black form λ_{max} = 633 nm [53].

The electrochromic behaviour of AIROFs is independent of the pH of the electrolyte solution [52] suggesting that neither protons nor hydroxide ions are involved in the electrochromic process. The mechanism for AIROFs is thus different from WO_3-based ECDs [53]; the anhydrous conditions that allow the tungsten trioxide to colour do not allow the AIROFs to operate. AIROFs are more stable in water than are tungsten trioxide ECDs provided the temperature remains low [44]: the bleached form of iridium oxide decomposes thermally above about 100 °C [54].

Beni, Rice and Shay [44] investigated the electrochromism of AIROF displays in both aqueous and non-aqueous electrolyte. They showed that some of the earlier experimental evidence is inconsistent with the hypothesis that the electrochromism of AIROFs is due to the extraction and insertion of hydroxide ions without proton insertion/extraction. They also showed that CN^- and F^- anions can be inserted during electrochromic operation in non-aqueous electrolytes.

The second method of film formation is reactive sputtering using an oxygen/argon mix (20 : 80). Such films are blue [46, 50] in the coloured state with λ_{max} = 610 nm. A denser SIROF, that may be electrochemically blackened, can be made using only oxygen as the flow-gas in the sputtering process. SIROFs have a complicated structure which, unlike AIROFs, is not macroscopically porous [55].

Black SIROFs are deposited as coloured films which can be decolourised by 85% on cycling, while blue SIROFs give superior films which may be transformed to a truly colourless state. In fact, blue SIROFs are very similar to AIROFs in being totally decolourisable. Furthermore, in terms of write–erase response times and absorbance spectra, blue SIROFs and AIROFs are again similar, cyclic voltammetry confirming the similarity [46]. Blue SIROFs have superior response times to black SIROFs, and a longer open-circuit memory. Beni and Shay [46] view the blue SIROFs as aesthetically the more pleasing.

Yoshino *et al.* have recently reverted to using anodically grown films of iridium oxide [56], deposited by a periodic reversal of the electrolysis potential, with an aqueous iridium sulphato complex as precursor. Such AIROFs are pronounced [56] to be better than SIROFs. The reliability of AIROFs is apparently variable [57].

Guterrez *et al.* [58] have investigated AIROFs using potential-modulated reflectance to tentatively assign the peaks in the cyclic voltammetry of anodic films of iridium oxide, and Ishihara *et al.* [59] have used iridium oxide in a solid-state device using reduced chromium oxide as the source of protons migrating into the electrochrome layer.

4.5 Molybdenum Trioxide

Films of molybdenum trioxide are often formed by anodic oxidation of molybdenum metal in acetic acid [60], or by vacuum evaporation [61–63]. Films may also be deposited electrochemically from an aqueous solution prepared by dissolution of molybdenum metal in hydrogen peroxide [64, 65]. Electrochromism is also evinced by solid phosphomolybdic acid [66].

Electroreduction of MoO_3 yields a 'bronze' :

$$\underset{\text{(colourless)}}{MoO_3 + x(M^+ + e^-)} \quad \rightarrow \quad \underset{\text{(blue)}}{M_x \overset{VI}{Mo}_{(1-x)} \overset{V}{Mo}_x O_3} \tag{4.7}$$

with the distorted rhenium trioxide structure shown in Fig. 4.2.

The product has a wavelength maximum centred at 770 nm [61]. This band is clearly not of simple origin [61], but comprises a group of discrete bands having maxima at around 500 nm, 625 nm, and 770 nm. The absorption edge of MoO_3 occurs at 385 nm [62], but shifts to ≈ 390 nm for the coloured reduced film [62]. The colouration efficiency for the molybdenum oxide bronze is therefore slightly greater than for the tungsten trioxide bronze since the absorption envelope coincides more closely with the visible region of the spectrum.

The coloured (dark-blue) form of the electrochrome is generated by simultaneous reduction of, and proton injection into, the MoO_3. Ord and co-workers have performed ellipsometric studies the proton injection into MoO_3 thin films [60, 67]. Their work is best interpreted as showing two different insertion sites for the (extremely mobile) hydrogen ion within the reduced film. There is a readily observed and well defined boundary between the oxidised and reduced regions within the oxide, perhaps in contrast to WO_3, see below. Ord *et al.* conclude that hydrogen insertion proceeds without the occurrence of major structural rearrangement in the bulk of the oxide film [68].

Evaporated MoO_3 films are colourless when deposited, but give an ESR signal characteristic of Mo^V at $g = 1.924$ [61]. X-ray photoelectron spectroscopy by Colton, Guzman and Rabalais [62] shows that the colour in the reduced state of MoO_3 is attributable to an intervalence transition between Mo^V and Mo^{VI} in the solid molybdenum bronze.

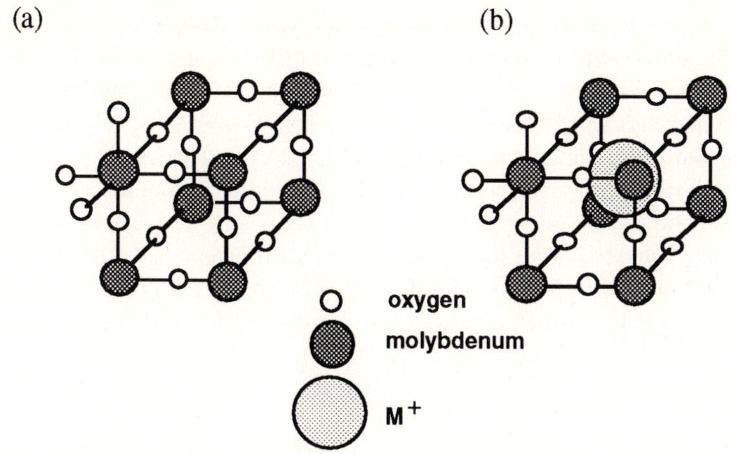

Fig. 4.2 Crystal structures (a) of molybdenum trioxide and (b) the perovskite 'bronze' product of reduction, M_xMoO_3.

Molybdenum bronzes show an improved open-circuit memory *cf.* tungsten bronzes, with H_xMoO_3 films oxidising more slowly than do H_xWO_3 films with similar values of x [69, 70]. Also, protons enter molybdenum oxide to form a bronze at potentials more anodic than about +0.4 V (*vs* the SHE) [70], leaving a colouration range of about 0.4 V before hydrogen gas evolves; the range for WO_3 is 0.5 V. It is claimed that the intensity of electrochromically generated colour may be enhanced by UV irradiation during colouration [71].

4.6 Nickel Oxide

Hydrated nickel oxide (also called 'hydroxide' or 'oxyhydroxide') undergoes electrochromism involving a colourless reduced form and a dark brown oxidised form [69]:

$$\underset{\text{(colourless)}}{NiO_x H_y} \rightarrow \underset{\text{(brown/black)}}{[\overset{II}{Ni}_{(1-z)}\overset{III}{Ni}_z]O_x H_{(y-z)}} + zH^+ + ze^- \tag{4.8}$$

Equation 4.8 is not consistent with all the experimental evidence available; for example, the ^{15}N nuclear reaction analysis method of Granqvist and Svensson [72] suggests that the reaction proceeds via proton incorporation. Here the sample is bombarded with high energy ^{15}N ions causing the nuclear reaction,

$$^{15}N + {}^{1}H \rightarrow {}^{12}C + {}^{4}He \tag{4.9}$$

The carbon-12 is formed in an excited nuclear state, which relaxes with the emission of gamma-rays, which, when counted, establish the hydrogen concentration within the sample. This technique does, however, cause localised heating of the sample and cannot therefore be classed as wholly non-destructive.

The problem of mass balance in the equation 4.8 is considered in depth by Bange *et al.* [73]. Nickel oxide films are typically amorphous, or microcrystalline to some slight extent, as determined by X-ray diffraction. Such films may be prepared by spraying aqueous nickel chloride [74] or nickel nitrate in aqueous butanol on to hot ITO [75]. A more common method of forming nickel oxide is RF-sputtering [71, 76–80] using an argon/oxygen gaseous mixture. Such films are deposited in their coloured state, but may be completely decolourised after only a few write–erase cycles [81] to become colourless or pale green. The colouration efficiency of RF-sputtered nickel oxide [81] is $-36\ cm^2\ C^{-1}$.

Films may also be prepared by DC sputtering [82, 83], although the electrochromic properties of such films are easily harmed by warming [83]. Vacuum deposition of nickel oxide (either thermal evaporation or electron-beam sputtering) is difficult.

An alternative method of preparing NiO_x films is electrodeposition, for example using an OTE immersed in (slightly alkaline) aqueous nickel nitrate [84–87], or using nickel sulphate solutions [88, 89]. Films may be grown by cathodising or cycling the potential [90]. Such films have colouration efficiencies as high as $-50\ cm^2\ C^{-1}$ [85] but a rather poor cycle life [91]. Thermal hydrolysis of either aqueous $[Ni(NH_3)_6]^{2+}$ or $NiSO_4$–urea mixtures is reported to form electrochromic films of NiO_x [92].

A study of the colouration process by Raman spectroscopy [93] shows that formation of defective crystal structure appear to be a prerequisite for electrochromism. The reduced form of the oxide contains some Ni^{3+}, and water is trapped at the defects. In separate studies by Corrigan *et al.* [94], evidence from both electrochemistry and spectroscopy was found for the presence of quadrivalent nickel in the oxidised form of nickel oxide. A SIMS study has detected phase changes that occur concurrently with ion insertion [95].

The refractive indices of $Ni(OH)_2$ films have been measured by Ord [96], and by Visscher and Barendrecht [84].

Most prototype electrochromic devices containing hydrated nickel oxide employ the film as the secondary electrode, for use in a complementary electrochromic sense, usually with tungsten trioxide [76, 77] as the primary electrochrome, but ECDs may also be constructed with $Ni(OH)_2$ as the primary [69, 85, 88]. Such an ECD has several advantages over more popular electrochromes like WO_3: the films have excellent durability even in water, high contrast ratios [69] of as much as 70:1 but, importantly, the nickel oxide is inexpensive.

Solid state ECDs utilising electrochromic nickel oxide have response times of 1 second or so [69], but no τ values are quoted for sputtered films in liquid electrolytes. A response time of < 1.0 s is quoted, however, for an electrodeposited film [85], the kinetics of colouration being controlled by ionic motion through the electrochrome [97].

4.7 Tungsten Trioxide

It was the discovery by Deb in 1969 [98] that thin films of tungsten trioxide exhibited electrochromism that initiated the serious quest for an electrochromic device. The mechanism for colouration is understood to involve the dual uptake of an electron and a counter cation, by which reductive process a series of tungsten 'bronze' compounds are formed (equation 4.10)

$$\underset{\text{(very pale yellow)}}{WO_3 + x(M^+ + e^-)} \rightarrow \underset{\text{(blue)}}{M_x\overset{VI}{W}_{(1-x)}\overset{V}{W}_x O_3} \tag{4.10}$$

M here may be either a metal (usually lithium) or hydrogen.

'Electrochromic Displays based on WO_3' by Faughnan and Crandall [99] (1980) and 'Electrochromic Tungsten-oxide Based Thin Films: Physics, Chemistry and Technology' by Granqvist [100] (1993) are both good introductions to the electrochromism of WO_3. Device-led research into ECDs has usually concentrated on tungsten trioxide as the electrochrome, in for example, 'smart windows' [101, 102]. WO_3 has been proposed for use in for alphanumeric watch-display characters [103], electrochromic mirrors [104–108] and other miscellaneous displays [109–117].

4.7.1 Operation of WO_3 ECDs

Following Deb's 1969 electrochromic experiments on solid WO_3, in 1975 significant progress ensued when Faughnan and Crandall [118] made a device with WO_3 in contact with liquid electrolyte. This ECD worked well at short times, but failed rapidly owing to film dissolution in the H_2SO_4 solution employed [119].

Film dissolution has been prevented by two different approaches. The first is the use of non-aqueous acidic solutions, for example, anhydrous perchloric acid in DMSO [53]. The effect of steadily drying the electrolyte has often been studied [120–123], while Randin [119] has reviewed the stability of WO_3 films in liquid electrolytes. Reichman and Bard's [124] study of the electrochromic processes of WO_3, using samples prepared by either anodic oxidation of tungsten metal, or by vacuum evaporation onto ITO, showed the electrochromic response time τ to be faster with the anodically grown material, which is porous. The value of τ was a function of the water content and film porosity, a greater porosity or water content enhancing τ ; unfortunately, such conditions produce films which are particularly susceptible to film dissolution. Reaction of acid with WO_3 prepared by anodising W metal is first order with respect to acid [125].

A different approach for inhibiting dissolution avoids the use of acid; rather, a non-protonic (alkali-metal) cation, usually lithium, is inserted into the tungsten trioxide.

Examples [122] include films of WO_3 immersed in dry propylene carbonate with $LiClO_4$ or $LiAsF_6$ as electrolyte.

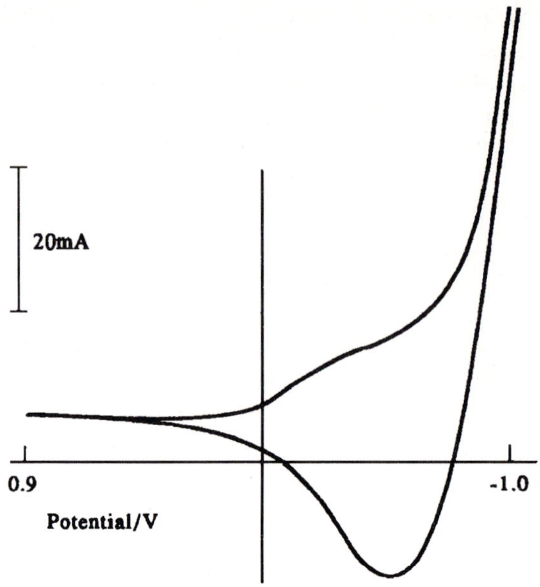

Fig 4.3 Cyclic voltammogram of electrodeposited WO_3 on platinum, in 0.1 M aq. HCl as electrolyte. The scan rate was 50 mV s^{-1}. (Figure reprinted from P.M.S. Monk and S.L. Chester, Electrochim. Acta, 38 (1993) 1521, copyright (1993), with permission from Elsevier Science Ltd., The Boulevard, Langford Lane, Kidlington, OX5 1GB, UK.)

Alternative thin-film WO_3 devices incorporate solid polymer electrolytes [126, 127]. Such cells have slower response times and also a poorer open-circuit memory, although Tell [128] made such a solid-state ECD from phosphotungstic acid claiming a τ of 10 ms (for an unspecified change in absorbance). Such liquid-free devices are preferred for their robustness.

Babinec [120a] has studied the colouration reaction using an electrochemical quartz-crystal microbalance. The insertion reaction is said to be complicated, even depending on the deposition rate employed in forming the electrochromic layer.

WO_3 in aqueous acidic electrolytes is more durable if the electrochrome–electrolyte interface is protected with a very thin layer of Nafion® [129], although charge transport through this layer is quite slow.

4.7.2 Structure, Preparation and Diffusion Characteristics

The tungsten trioxide lattice has the distorted rhenium trioxide structure (Fig. 4.2 a) and the bronze product of equation (4.10) is a perovskite (Fig. 4.2 b). The slight structural changes to the WO_3 lattice occurring on reduction are a likely cause of physical stresses: Green states that evaporated films expand by *ca.* 6% on colouration [130]. A spectroelectrochemical study of WO_3 colouration (measuring transmittance in tandem with potential ramps) [131] established that electrochemical colouration was accompanied by two optical absorption steps which are due to either a structural rearrangement, or follow some charge diffusion phenomenon.

Tungsten trioxide (as a thin film) exists either in an amorphous or a micro-crystalline state, a-WO_3 or c-WO_3 respectively. The former is commonly formed by thermal evaporation *in vacuo*, while the latter is produced by sputtering W in an oxygen containing atmosphere, or by thermal annealing of a-WO_3. The method of film preparation dictates the morphology of the tungsten trioxide formed. For example, Deb's [132] evaporated WO_3 was shown to be amorphous by X-ray diffraction, but WO_3 films prepared by RF-sputtering are partially crystalline [81]. In a brief review, Whittingham [133] has discussed the structures and thermodynamics of phases formed during electrolysis in WO_3.

Crystalline WO_3 is not only optically absorbent, but also partially reflective. Svensson and Granqvist [134] have analysed the reflectance in terms of the optical constants, their results illustrating the relationship between crystallinity in the film and the degree of reflectance observed. In fact, most of the physical and optical properties of WO_3 and its reduction products are highly sensitive to the method of film preparation. Other methods of film preparation commonly used (see chap. 3) are RF-sputtering [81, 130, 135, 136] and pyrolysis of $W(CO)_6$ in a stream of oxygen gas (i.e. chemical vapour deposition, CVD) [137–140]. Vacuum-evaporated WO_3 is discussed in references [120a, 124, 132, 141].

In a different approach, Livage [142] applied a gel containing colloidal hydrogen tungstate to an OTE. Layers of WO_3 were formed as a result of thermal curing. The sol-gel method is particularly suited to the preparation of large-area ECDs, for example, for fabricating electrochromic windows. Response times of 40 s are reported [142]. An open circuit memory in excess of six months has been cited [143, 144]. Tungsten alkoxides as sol-gel precursors yield electrochromic layers [145]. WO_3 has also been prepared by a dip-coating procedure [146, 147].

Table 4.1 (a): Diffusion Coefficients D of Lithium Ions in WO_3, as Li_xWO_3.

Morphology†	x	$D/cm^2\ s^{-1}$	Notes	Ref.
WO_3	0	10^{-9}	*, a, b	[162]
a-WO_3	0.097	2.4×10^{-12}	*, b, c	[158]
a-WO_3	0.138	4.9×10^{-12}	*, b, c	[158]
a-WO_3	0.170	1.5×10^{-11}	*, b, c	[158]
a-WO_3	0.201	2.6×10^{-11}	*, b, c	[158]
a-WO_3	0.260	2.8×10^{-11}	*, b, c	[158]
WO_3	—	1.6×10^{-12}	*, b, d	[163]
WO_3	—	10^{-11}	*, b, d	[164]
WO_3	0	2.1×10^{-11}	*, a, b	[165]
WO_3	—	4×10^{-13}	*, d, e	[130]
WO_3	—	2×10^{-11}	*, e	[166]
WO_3	—	5×10^{-13}	f	[177]

Table 4.1 (b): Diffusion Coefficients of Protons in WO_3.

Morphology†	$D/cm^2\ s^{-1}$	Notes	Ref.
a-WO_3	2×10^{-8}	*, c, d	[154]
c-WO_3	$10^{-14} - 10^{-11}$	*, a, d, h	[168]
WO_3	8×10^{-10}	*, d	[164]
WO_3	4×10^{-10}	b, c	[169]
WO_3	$2 \times 10^{-10} - 10^{-9}$	*, c, i	[170]
WO_3	5×10^{-8}	*, e, i	[170]
WO_3	2×10^{-11}	*, b, c, h	[164]
WO_3	10^{-12}	g	[171]
WO_3	$10^{-12} - 10^{-11}$	h	[155]

† Insertion coefficient $x \approx 0$ in all cases

Key: * Thin film. [a] Sputtered film. [b] Impedance measurement. [c] Thermally-evaporated sample. [d] Chronoamperometric measurement; [e] Electrochemical film preparation. [f] Film prepared using acidified tungstate solution as sol-gel intermediate. [g] Single crystal. [h] Thermally-annealed sample. [i] Measured using cyclic voltammetry.

There have recently been many studies of WO_3 prepared by electrodeposition [148–151], the voltammetry solution comprising either the oxotungstate anion $[(O_2)_2(O){-}W{-}(O){-}(O_2)_2)]^{2-}$ [152] (formed by dissolving powdered tungsten metal in hydrogen peroxide [148–150]) or tungsten carboxylates [151]. Such films on ITO or Pt sometimes appear to be slightly gelatinous and are essentially amorphous as shown by XRD. Cation diffusion through WO_3 has received particular attention, using hydrogen ions [153–156], deuterium ion [157] and lithium ion [158–160]. Heat curing (that is, annealing) of WO_3 is said to decrease the response time [159], the effect being due to the increased proportion of the WO_3 which is crystalline; the temperature at which the (endothermic [119]) amorphous-to-crystalline transition occurs is *ca.* 90 °C [161]. The rates of charge transport in electrochromic WO_3 films is the subject of a brief review by Goldner [172]. For example, the chemical diffusion coefficient (i.e. accompanied by gross composition change) of the guest ion within the host film can vary from 10^{-13} cm^2 s^{-1} (lithium in *c*-WO_3 [130]) or 10^{-7} to 10^{-6} cm^2 s^{-1} (for lithium in $K_{0.33}WO_3$ along a particular crystallographic axis (the *a*–axis) [173]). Table 4.1 contains many values of the diffusion coefficient for motion of either Li^+ or H^+ through WO_3, which vary owing to variations in structure crystallinity, water content and the like. (The composition change also impinges on values obtained by different techniques).

4.7.3 Spectroscopic and Optical Effects

Pure bulk tungsten trioxide is pale yellow. The colour of the WO_3 actually deposited depends on the method of film preparation, thin films sometimes showing a slight pale-blue tint owing to oxygen deficiency in a sub-stoichiometric oxide $WO_{(3-y)}$. Deb [132] gives $y = 0.03$ but Bohnke and Bohnke [159] quote 0.3. In fact, the extent of oxygen deficiency depends on the temperature of the evaporation boat and/or of the substrate target [174]. It is found that the diffusion coefficient of lithium ions being inserted into RF-sputtered WO_3 decreases as the extent of oxygen deficiency increases [175].

Electrochemically reduced tungsten trioxide films have an intense blue colour, the spectrum exhibiting a broad structureless band peaking in the near infra-red (see Fig. 4.4). In transmission, the electrochromic transition is effectively colourless to blue at low x ($x \leq 0.2$). At higher x, insertion irreversibly forms a metallic 'bronze' which is red or golden in colour.

The product of reduction, M_xWO_3, that is, $M_xW^{VI}_{(1-x)}W^{V}_xO_3$ (eq. 4.10), when viewed by reflected light, has a colour which is a function of x (itself proportional to the charge injected) according to scheme (4.1) [133, 176]:

x =	0.1	0.2–0.4	0.6	0.7	0.8–1.0
	grey	*blue*	*purple*	*brick-red*	*golden-bronze*

Scheme 4.1

The maximum of the spectral peak is morphology-dependent, and perhaps sensitive to water content. λ_{max} shifts from 900 nm in reduced films H_xWO_3 which are amorphous and hydrated [70, 109, 177–179], to longer wavelengths in polycrystalline [179] materials: λ_{max} = 1300 nm for samples which contain an average grain size of 250 Å [80, 177, 180]. The colour of the bronze is independent of the cation used during reduction [62, 177, 181], be it M = H^+, Li^+, Na^+, K^+, Cs^+, Ag^+ or Mg^{2+} (stoichiometrically as $^1/_2$ Mg^{2+}). Most of these cations cannot be inserted reversibly into WO_3, only H^+ and Li^+ being readily retrieved following insertion. Indeed, a quartz-crystal microbalance study suggests that the colouration usually attributable to Li^+ is in fact caused by proton insertion followed by swapping with Li^+ at longer times [122].

The nonmetal-to-metal transition in H_xWO_3 occurs at a critical composition x of $x_c = 0.32$, determined for amorphous H_xWO_3 by conductivity measurement [181]. Below x_c, the bronze is commonly held to be a mixed-valence species [182], and belongs to Group II (involving moderate delocalisation) of the Robin and Day [3] classification scheme cited in section 4.1. H_xWO_3 with $x > x_c$ is metallic and completely delocalised (i.e. now within the Robin and Day Group IIIB). It is the unbound electron plasma in metallic WO_3 bronzes which confers the reflectivity [153, 183–185]. Schirmer *et al.* [177, 186] had earlier dismissed the existence of a Drude-type absorption (i.e. due to free electrons) in amorphous WO_3.

There is still some controversy concerning the cause of the blue colour of tungsten bronzes at compositions below x_c. Deb [132] suggests the absorption is due to F-centre-like colour centres, localised at oxygen vacancies within the WO_3 sub-lattice. Chang *et al.* [187] state that the origin of the blue colour is electrochemical oxygen extraction, the coloured product being sub-stoichiometric $WO_{(3-y)}$; Faughnan and Crandall [118] propose a model where injected electrons are predominantly localised on W^V atoms [188], a $W^V \rightarrow W^{VI}$ intervalence transition being responsible for the colour. Faughnan's model is clearly right. The electron localisation and the accompanying lattice distortion around the W^V may be treated as a bound small polaron [62, 132, 174, 188–191].

Pfifer and Sichel [192], who studied the ESR spectrum of H_x WO_3 (at low x), could find no evidence for the presence of unpaired W^V electrons. A likely interpretation of this observation is that ground-state electrons form paired rather than single spins, probably at adjacent W^V sites [190, 191].

Graphs of absorbance for electrogenerated M_xWO_3 against the quantity of charge consumed (eq. 4.10) in forming the bronzes, are akin to a Beer-Lambert law plot of absorbance versus concentration, since each electron transferred generates a colour centre. Such a graph, if linear, implies the absence of any electrochemical side reactions. The gradient of this graph for a sample of unit area is the colouration efficiency η (eq. 1.4). A Beer-Lambert law plot for thin-film WO_3 is only linear for small values of x ($0 < x \leq 0.03$ [70, 118] or 0.04 [193]); this result applies both for the insertion of protons [70, 118, 193] and sodium ions [194] in evaporated (amorphous) WO_3 film [118, 193]. A

Beer-Lambert law plot for lithium ion insertion into evaporated (amorphous) WO_3 is linear to larger x, but has a smaller gradient (i.e. smaller η) [109].

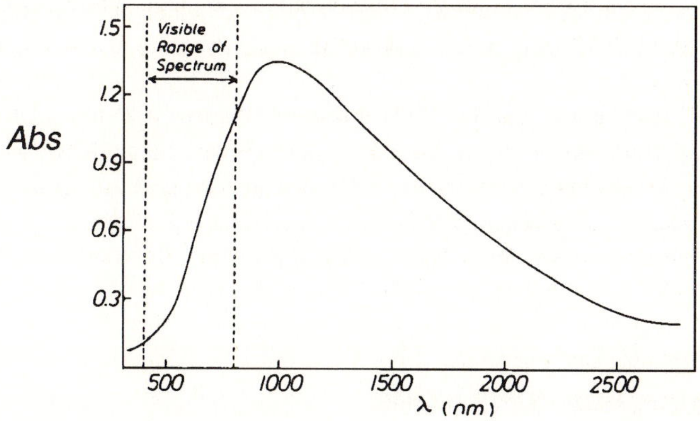

Fig. 4.4 UV-vis spectrum of tungsten bronze with composition $H_{0.17}WO_3$ on ITO. (Figure reproduced from ref. [193a] with the permission of Butterworths.)

While the colouration efficiency for Li^+ insertion is independent of x (giving a linear Beer–Lambert plot) until x is quite large [194], for H^+ or Na^+, however, the Beer-Lambert gradient decreases with increasing x (that is, η decreases as x increases). Such non-linearity is not due to competing electrochemical side-reactions [70]; rather, it is thought to be due to either a decrease in the oscillator strength [193] per electron, or a broadening of the envelope of the absorption band.

A different behaviour is exhibited by films of polycrystalline WO_3, prepared for example by RF-sputtering, or by high temperature annealing of amorphous WO_3: at low x, the Beer–Lambert plot is linear (but of low gradient) but η *increases* with an increase in x [109, 195]: this effect may be due to specular reflection, that is, a not wholly absorptive phenomenon. Sputtered films prepared using a target of tungsten metal yield films which evince a different Beer–Lambert behaviour to sputtered films made using a WO_3 target [196].

In the treatment of Duffy *et al.* [193] (who used evaporated WO_3) four linear regions are identified in the Beer–Lambert plot, each with a different apparent extinction coefficient. It is emphasised that reduction is not envisaged to proceed fitfully, with sudden mechanism changes at discrete values of x: structural effects or oscillator strength or bandwidth are implicated.

For thin films of WO_3, prepared by chemical vapour deposition (CVD) [137–139], Beer–Lambert plots are linear for H^+ or Li^+ when the insertion coefficient is low. However, η decreases at higher x, although the value of x at which curvature begins were not stated.

Ellipsometric studies by Ord [197] of thin-film WO_3 (grown anodically) show little optical hysteresis associated with colouration, provided the reductive current is applied for a limited duration: films return to their original thicknesses and refractive indices. Colour cycles of longer duration, however, reach a point at which further colouration is accompanied by film dissolution (*cf.* comments in section 1.4.6 and above, concerning cycle lives).

Also [197], the optical data for WO_3 grown anodically on W metal fit a model in which the colouring process takes place by a progressive change throughout the film, rather than by the movement of an interface that separates coloured and uncoloured regions of the material, and a more recent study [198] concludes that a 'substantial' fraction of the H^+ inserted during colouration cycles is still retained within the film when bleaching is complete.

4.8 Vanadium Pentoxide

Vanadium pentoxide films may be prepared by evaporation *in vacuo* [62, 165, 199] or more commonly by reactive RF-sputtering [200–205], using a high pressure of oxygen and a target of vanadium metal. Spin-coating has also been used [206, 207].

Films deposited by evaporation are amorphous [62], while films of sputtered V_2O_5 are crystalline [202, 204] although X-ray diffraction suggests the extent of crystallinity to be marginal [204]. Heating to 180 °C increases the crystallinity [208]. Films formed by either method show a characteristic yellow/brown aspect, attributable to tailing from the UV of the optical band edge into the visible region.

The electrochromic reaction is

$$\underset{\text{(very pale blue)}}{M_xV_2O_5} \quad \rightarrow \quad \underset{\text{(brown/yellow)}}{V_2O_5 + x\,(M^+ + e^-)} \tag{4.11}$$

where M^+ is usually Li^+. The electrochromism of thin-film V_2O_5 was first mentioned in 1977 by Gavrilyuk and Chudnovski [209], who prepared samples by thermal evaporation. Since thin-film V_2O_5 dissolves readily in dilute acid, alternative electrolytes have been used, for example, distilled water [209], LiCl in anhydrous methanol [210] or $LiClO_4$ in propylene carbonate [200, 201, 204].

The electrogenerated colour is blue-green for evaporated films [211] at low insertion levels, going via dark-blue to black at higher insertion levels [209]; the colour changes from purple to grey if films are sputtered [70]. Rauh *et al.* [200] state that for certain film thicknesses V_2O_5 is colourless between brown and pale-blue states.

Electroreduction of the film causes the absorption spectrum to change greatly, the yellow colour being completely removed and a broad but relatively weak band developing in the near IR [200] centred on 1100 nm. Also, the optical band edge shifts to higher energy, even for low insertion levels [70].

Cyclic voltammetry of sputtered V_2O_5, as a thin film supported on an OTE in a lithium-containing propylene carbonate electrolyte, shows two well-defined quasi-reversible redox couples with anodic peaks at 3.26 and 3.45 V, and cathodic peaks at 3.14 and 3.36 V relative to the Li/Li^+ couple in propylene carbonate [201]. These two pairs of peaks may correspond to the two phases of $Li_xV_2O_5$ identified by Dickens and Reynolds [212].

Vanadium pentoxide itself has a distorted structure in which the nominally octahedral vanadium is almost tetragonal bipyramidal, with one distant oxygen [213]. Reductive injection of lithium ion into V_2O_5 forms $Li_xV_2O_5$. The $Li_xV_2O_5$ (of $x < 0.2$) prepared by sputtering is the a-phase, which is not readily distinguishable from the starting pentoxide [200]. At higher injection levels ($0.3 < x < 0.7$), the crystalline form of the oxide is ε–$Li_xV_2O_5$ [200], as identified by Hub et al. [210] and Murphy et al. [214]. The ε phase of $Li_xV_2O_5$ in V_2O_5 thin films accompanies the electrochromic colour change. a–$Li_xV_2O_5$ from the unlithiated oxide is also formed and contributes an additional slight change in absorbance [200]. Since several species are participating in the spectrum of the bronze, spectral regions following the Beer-Lambert law cannot be identified readily [201].

The absorption bands formed on reduction are generally considered to be too weak to imply the formation of any intervalence species. From X-ray photoelectron spectroscopy, Fujita *et al.* [199], assigned the colour change in evaporated films incorporating lithium to the formation of VO_2 in the V_2O_5. Colton, Guzman and Rabalais, [62], using the same technique, did infer a weak charge-transfer transition between the oxygen $2p$ and vanadium $3d$ states, provided that the sample is all V^{IV}.

Ellipsometric studies [215, 216] of evaporated V_2O_5 showed, in common with MoO_3 (but unlike WO_3), that a well-defined boundary is formed between the coloured and bleached phases during cycling. This boundary moves into the film from the film-electrolyte interface [215] during the bleaching and colouration processes. Higher fields are required for bleaching than for colouring [215]. Scarminio *et al.* have monitored the stress induced in V_2O_5 on lithium insertion [217].

Since the electrochromic colours of V_2O_5 films are yellow and blue, the *CR* for such films is not great, hence the system is currently being investigated for possible use in ECDs as the secondary electrochromic layer for counter-electrode use [165, 200, 201].

Thin films of lithium vanadate ($LiVO_2$) are also electrochromic [218].

4.9 Other Metal Oxides

4.9.1 Cerium Oxide

Cerium oxide is electrochromic [219, 220]:

$$\underset{\text{(yellow)}}{CeO_2 + x(M^+ + e^-)} \rightarrow \underset{\text{(very pale blue)}}{M_xCeO_2} \tag{4.12}$$

Since the colour change is not intense, and the movement of ionic charge through is the oxide is slow [221], this material is unlikely to have any electrochromic applications except as a secondary electrochrome.

4.9.2 Iron Oxide

Although films of iron oxide are electrochromic [222, 223], the slight electrochemical irreversibility they evince will probably preclude their utilisation as viable electrochromes. For example, yellow/green films form on the surface of iron electrodes anodised while immersed in 0.1 M NaOH [223]. This coloured material is thought to be hydrated $Fe^{III}O{\cdot}OH$. The film becomes transparent at cathodic potentials as hydrated $Fe(OH)_2$ is formed.

4.9.3 Manganese Oxide

Electrochromic films of manganese oxide are generated similarly by anodising Mn metal in alkaline solution [224]. The film had two readily formed colours, being yellow at low potentials and red/brown at higher potentials; the film appears black if thick. The yellow film was thought to consist of hydrated MnO_2 while the red/brown film probably has intervalence character containing for example Mn_3O_4. The electrochromic process is complicated but appears to involve proton uptake:

$$\underset{\text{(yellow)}}{MnO_2 + z\,e^- + zH^+} \rightarrow \underset{\text{(brown)}}{MnO_{(2-z)}(OH)_z} \tag{4.13}$$

Alternative manganese oxide electrochromes may be prepared by electrodeposition, for example using manganous sulphate (of pH 9.2) and an SnO_2 OTE as the conducting substrate [225, 226].The coloured brown form of the oxide prevails at potentials more anodic than 0.8 V while the bleached yellow form occurs below 0.0 V. The electrochemistry of electrochromic MnO_2 films is complicated since deep cycles cause a loss of electrochromic activity. The intense electrochromic colour of the brown material is attributable to an optical transition between Mn^{3+}/Mn^{4+} centres [226]; in the UV, η is reportedly $-140\ cm^2\ C^{-1}$ at $\lambda = 350$ nm. Films of MnO_2 have also been produced by RF-sputtering [229] and electron-beam sputtering [230, 231]. A recent Raman spectroscopic investigation of electrodeposited MnO_x films concluded that films were unsuitable for electrochromic applications owing to poor reversibility [227, 228].

4.9.4 Niobium Pentoxide

Amorphous niobium pentoxide has been incorporated in an ECD [232] with aqueous HF or H_3PO_4 as the electrolytes, or with $LiClO_4$ in propylene carbonate [233, 234]. The oxide deposited is white and the colour of bronzes formed on reduction, with x small at *ca.* –0.6 V, is pale blue [31]:

$$\underset{\text{(colourless)}}{Nb_2O_5 + x\,(M^+ + e^-)} \rightarrow \underset{\text{(pale blue)}}{M_xNb_2O_5} \tag{4.14}$$

Since the colouration efficiency η for the oxide is small and negative (see table 1.2) films of niobium pentoxide are best used as secondary electrochromes [233, 234].

Nb_2O_5 films may also be prepared by DC-magnetron sputtering of Nb nitride [235] or thermal oxidation of Nb metal [236]. Crayston and Lee [237] prepared films of Nb_2O_5 using a sol-gel intermediate itself prepared by alcoholysis of $NbCl_5$ spin-coated onto ITO and subsequent dipping into aqueous acid. Films were relatively unstable in liquid electrolytes such as $LiClO_4$–MeCN, but were durable in a siloxane composite. Diffusion coefficients of Li^+ through this Nb_2O_5 were rather small [237].

4.9.5 Palladium Oxide

Palladium oxide is electrochromic, existing below 1.2 V as a yellow oxide, becoming ruddy brown in hue if the potential is increased to about 1.6 V [238].

4.9.6 Rhodium Dioxide

Metallic rhodium [239, 240] forms an electrochromic oxide coating when anodised in alkaline solution, although the mechanism of the colouration step comprises a series of complicated equilibria involving soluble intermediate(s). Unless the conditions for electrochromism can be optimised, this metal oxide system is not viable as an electrochromic species for ECD inclusion. The reaction for rhodium is said to be [240]

$$RhO_2{\bullet}2H_2O + H_2O + e^- \rightarrow {}^1/_2\,(Rh_2O_3{\bullet}5H_2O) + OH^- \qquad (4.15)$$

(yellow) (dark green)

with the oxidative electron-transfer reaction occurring at about 1.0 V. The expense of Rh (like Ru, following) limits the electrochromic usefulness.

4.9.7 Ruthenium Dioxide

As with rhodium, thin-film ruthenium oxide, generated by anodising metallic Ru in alkaline solution, changes colour electrochromically [241] but not very intensely:

$$RuO_2{\bullet}2H_2O + H_2O + e^- \rightarrow {}^1/_2\,(Ru_2O_3{\bullet}5H_2O) + OH^- \qquad (4.16)$$

(blue/brown) (black)

4.9.8 Titanium Oxide

Titanium oxide has a poor colouration efficiency (see table 1.2) but may be used in counter electrodes [219]. Dip coated samples have been made [242].

The rate of TiO_2 (anatase) reduction is controlled by ionic diffusion through the solid [243]. For example, ionic insertion into anatase (Li^+ from a $LiClO_4$/propylene carbonate electrolyte) is characterised by a diffusion coefficient of 10^{-10} cm^2 s^{-1} [244].

Using ellipsometry Ord *et al.* [245] have studied the electrochromism of titanium oxide grown anodically on Ti. Both reduction and oxidation proceed via movement of a phase boundary separating reduced and oxidised regions in the TiO_2. In aqueous electrolyte, the rate of movement is limited by competition between the electron-transfer and hydrogen-evolution rates.

4.10 Mixed Metal Oxides

Recently, many workers have prepared films of metal oxide containing other metal oxides. Such mixtures are often said to be 'doped'. The presence of even small amounts of a guest oxide within the electrochrome host can have profound effects on the spectroscopic characteristics of the material, its conductivity and the potential window available for electrochromic operation.

4.10.1 Cobalt Oxide Mixtures

Solely cobalt oxide electrochromes are considered in section (4.2). Electrochromic oxides have been grown on cobalt/nickel alloys [246]. Cobalt oxides doped with Cu, Ni, Mo, W and Zn have been prepared by electrodeposition from an aqueous solution containing equimolar cobalt and dopant cation [19, 247].

Incorporation of additional metal oxides greatly increases the colouration efficiency η of the cobalt oxide, and the product of reduction is more blue than for the pure CoO host. The films are also physically stronger. The diffusion coefficients D are generally much larger in mixed M/Co oxide films than in cobalt oxide alone [19, 247]. Notably, mixed-metal oxide electrochromes containing cobalt all colour cathodically while CoO itself colours anodically.

4.10.2 Molybdenum Trioxide Mixtures

Solely molybdenum trioxide electrochromes are considered in section (4.5). A film of the general formula $Mo_{(1-x)}W_xO_3$ is formed if molybdenum trioxide is co-evaporated *in vacuo* with tungsten trioxide. The wavelength maxima of mixed oxides following reduction are shifted, relative to the pure oxides, to higher energies [177, 248]. For such mixed-oxide films in the reduced state, the relationship between λ_{max} and the quantity of charge injected appears complicated [81] *cf.* WO_3, for which the value of λ_{max} is independent of the insertion coefficient; for H_xMoO_3 there is a slight (linear) dependence between the absorbance maxima and x, λ_{max} increasing as x decreases [81]. Reference [81] quotes various colouration efficiencies for such composite films, although no details of film preparation are given. For example, η = 65 cm^2 C^{-1} for H_xWO_3, 77 cm^2 C^{-1} for H_xMoO_3 and 110 cm^2 C^{-1} for $H_xW_{0.992}Mo_{0.008}O_3$, all measured at 700 nm. For a device prepared with the mixed film, the response time to produce a given contrast ratio will be correspondingly faster than for pure oxide films, and the energy consumption of mixed-oxide films will also be smaller. The observed decrease in electron mobility within mixed-metal oxide films [248] is not thought to be deleterious to device performance [70].

Mixed molybdenum-tungsten trioxide films may also be prepared by CVD [138] or electrodeposition [65]. It is interesting to note that electrodeposited films containing Cr or Fe, while exhibiting a rather poor contrast ratio, have a greatly extended potential window with respect to films containing no dopant, inhibiting the formation of molecular hydrogen when aqueous acidic solutions are used.

4.10.3 Nickel Oxide Mixtures

Solely nickel oxide electrochromes are considered in section (4.6). Corrigan *et al.* [86, 87] have reported the preparation of nickel oxide films in which other transition-metal cations are co-precipitated along with nickel during deposition. Using an aqueous alkaline solution of $Ni(NO_3)_2$, together with the relevant metal (also as the nitrate) in the ratio 10:1, Corrigan included the additional metal ions Ag, Cd, Ce, Co, Cr, Cu, Fe, La, Mg, Mn, Pb and Y [249, 250]. Films containing tungsten can also be formed [149, 150]. In all cases, the reduced films were essentially transparent, while the oxidised films exhibited intense, broad absorption bands throughout the visible region [86].

Such co-precipitation is stated to have a considerable effect on the switching speed of the electrochromic nickel couple [86], Ce, Cr and La improving the colouration rate, while Ce, Cr and Pb cause slower bleaching [86]. Films containing yttrium have very slow colouration times (*ca.* 10 s), and films containing silver exhibited complicated behaviour [86]. Significantly for electrochromic display applications, co-precipitation of cerium or lanthanum ion appears to improve film durability [86]. Also, the over-potentials of hydrogen and/or oxygen evolution (from electrolysis of the electrolyte) are increased for some co-precipitated films. Films of nickel oxide mixed with the oxides of manganese or niobium have also been studied [251].

4.10.4 Tungsten Trioxide Mixtures

Solely tungsten trioxide electrochromes are considered in section (4.7). Electrochromic films of WO_3 have often been doped with metals such as platinum and gold [252]. WO_3 has also been doped with the oxides of barium [253], cobalt or nickel [149, 150, 254], molybdenum [255], tin [256] and titanium [257, 258].

Recently, it has been shown that the oxides of Ag, Co, Cr, Cu, Fe, Mo, Ni, Ru or Zn can each be incorporated into a WO_3 matrix [150]. Films containing silver and copper are not very useful as they tend to form metallic products during reduction rather than yielding the desired doped-oxide product. Films containing Co, Ni or Zn were the most promising in terms of contrast ratio and durability, and protonic diffusion through the oxide was also rapid.

WO_3–TiO_2 films have also been made either by sputtering [259] or using sol-gel intermediates [259, 260].

Electrochromic $HNbWO_6$, in sulphuric acid has a similar transparent-to-blue electrochromic operation to WO_3 but with a superior stability to dissolution [261, 262].

4.10.5 Vanadium Oxide Mixtures

Solely vanadium oxide electrochromes are considered in section (4.8).Vanadium pentoxide containing copper, silver or gold has been formed by laying down alternate layers of the constituents during vacuum evaporation of thin-film samples [263]. Films containing gold are superior to Ag–V_2O_5 or Cu–V_2O_5 films. Au–V_2O_5 is deposited as a green material, the colour becoming yellow after calcination at > 300 °C.

The electrochromism reported [263] for Au–V_2O_5 has a possible colour change violet-to-green in the potential range −0.8 V to +1.2 V. A new red/violet colour was also observed at potentials below about −1.0 V.

Both Ag and Cu containing films were orange after calcination, on reduction Cu–V_2O_5 becoming dark brown, Ag–V_2O_5 turning blue/green.

Cr–V_2O_5 and Nb–V_2O_5 have also received attention [264] as have TiO_2–V_2O_5 films made by a sol-gel process [265] or by spin coating [266].

4.10.6 Miscellaneous Metal Oxide Mixtures

Electrochromic mixtures of cerium oxide with either titanium dioxide or zirconium dioxide have been prepared via sol-gel intermediates [267]. Similarly, a CeO_2-TiO_2 films may be made [220, 268], for example, by a dip-coating procedure [268]. Iridium-ruthenium coating electrodeposited on titanium are electrochromic [269].

4.10.7 Ternary Oxide Mixtures

All the oxides in section 4.10 above are binary. Reports of ternary oxides which are electrochromic are rare. Examples include amorphous $(Li_2B_4O_7)_{(1-x)}(WO_3)_x$ [270] or sintered $(WO_3)_x(Li_2O)_y(MO)_z$ (where M = Ce, Fe, Mn, Nb, Sb or V) [271], although the latter has a poor transmittance which may preclude all but reflective use [272]. Oxides of the type $M_x\,M'_y\,W_{(1-x-y)}O_3$ (M, M' = Co, Cr, Mo, Ni, Zn) have recently been prepared by electrodeposition [273] and show superior colouration efficiencies to any of the parent oxides alone. Interestingly, films are more durable and are much stronger physically when the mole fraction of tungsten is relatively small.

Germanates and stannates of cadmium doped with zinc are also electrochromic [274, 275]. The coloured forms of the oxides are apparently sensitive to air.

Electrochromic films containing four oxides have also been prepared by electro-deposition [273].

4.11 Metal Oxide–Organic Mixtures

A different class of mixture is seen when a metal oxide is dispersed in a conducting polymer. For example, tungsten trioxide within a polypyrrole [276, 277] or polyaniline matrix has been reported to be electrochromic [278–280].

References

[1] D.B. Brown (ed.), 'Mixed Valence Compounds' (NATO Conference), D. Reidel, London, 1980.

[2] K. Prassides (ed.), 'Mixed Valency Systems: Applications in Chemistry, Physics and Biology' (NATO Conference), Kluwer, London ,1991.

[3] M.B. Robin and P. Day, *Adv. Inorg. Chem. Radiochem.*, 10 (1967) 247.

[4] N.S. Hush, *Prog. Inorg. Chem.*, 8 (1967) 357.

[5] F.G.K. Baucke, J.A. Duffy and R.I. Smith, *Thin Solid Films*, 186 (1990) 47.

[6] F.A. Cotton and G. Wilkinson, 'Advanced Inorganic Chemistry', 4th edn., Wiley, 1980, p. 767.

[7] L.D. Burke and O.J. Murphy, *J. Electroanal. Chem.*, 109 (1980) 373.

[8] L.D. Burke and O.J. Murphy, *J. Electroanal. Chem.*, 112 (1980) 379.

[9] W.K. Behl and J.E. Toni, *J. Electrochem. Soc.*, 31 (1971) 63.

[10] P. Benson, G.W.D. Briggs and W.F.K. Wynne-Jones, *Electrochim. Acta*, 9 (1964) 275.

[11] A. Bewick, C. Gutiérrez and G. Larramona, *J. Electroanal. Chem.*, 333 (1992) 165.

[12] G.Wei, T.E. Haas and R.B. Goldner, *Solid State Ionics*, 58 (1992) 115.

[13] R.B. Goldner, T.E. Haas, and G. Wei, *Proc. Electrochem. Soc.*, 90-2 (1990) 80.

[14] R.B. Goldner, F.O. Arntz, G. Berera, T.E. Haas, G. Wei, K.K. Wong and P. Yu, *Proc. S.P.I.E.*, 1536 (1991) 63.

[15] R.B. Goldner, F.O. Arntz, G. Berera, T.E. Haas, G. Wei, K.K. Wong and P. Yu, *Solid State Ionics*, 53-56 (1992) 617.

[16] G. Wei, Ph.D thesis, 1991, Tufts University (USA), *Diss. Abstr. Int. B.*, 52 (1991) 2247, cited in *Chem. Abstr.* 116: 116,951*d*.

[17] A. Gorenstein, C.N. Polo da Fonseca and R. Torresi, *Proc. S.P.I.E.*, 1536 (1991) 104.

[18] M. Bendert and D.A. Corrigan, *J. Electrochem. Soc.*, 136 (1989) 1369.

[19] P.M.S. Monk, S.L. Chester and D.S. Higham, *Proc. Electrochem. Soc.*, 94–2 (1994) 100.

[20] R.W. Murray, W.R. Heineman and C.W. O'Dom, *Anal. Chem.*, 39 (1967) 1666.

[21] T. Kuwana and W.R. Heineman, *Acc. Chem. Res.*, 9 (1976) 241.

[22] K.L. Chopra, S. Major and D.K. Pandya, *Thin Solid Films*, 102 (1983) 1.
[23] N.R. Lynam, *Proc. Electrochem. Soc.*, 90–2 (1990) 201.
[24] S.J. Golden and B.C.H. Steele, *Solid State Ionics,* 28-30 (1988) 1733.
[25] B.C.H. Steele and S. Golden, *Appl. Phys. Lett.*, 59 (1991) 2357.
[26] S.J. Golden and B.C.H. Steele, *Mater. Res. Soc. Symp. Proc.*, 293 (1993) 169, cited in *Chem. Abstr.* 120: 90,240*b*.
[27] N.R. Armstrong, A.W.C. Lin, M. Fujihira and T. Kuwana, *Anal. Chem.*, 48 (1976) 741.
[28] S.F. Cogan, T.D. Plante, E.J. Anderson and R.D. Rauh, *Appl. Optics,* 24 (1985) 2282.
[29] R.B. Goldner, G. Foley, E.L. Goldner, P. Norton, K. Wong, T.E. Haas, G. Seward and R. Chapman, *Appl. Optics,* 24 (1985) 2283.
[30] P.C. Yu, T.E. Haas, R.B. Goldner and S.F. Cogan, *Mat. Res. Soc. Symp. Proc.*, 210 (1991) 63, cited in *Chem. Abstr.* 115: 242,820*c*.
[31] P.C. Yu, Ph.D Thesis, 1991, Tufts University (USA). *Diss. Abstr. Int. B.*, 52 (1991) 2025, cited in *Chem. Abstr.* 116: 70,431*d*.
[32] J.S.E.M. Svensson and C.G. Granqvist, *Appl. Optics,* 24 (1985) 2284.
[33] S.J. Jiang and C.G. Granqvist, *Proc. S.P.I.E.*, 562 (1985) 128.
[34] Z. Ovadyahu, B. Ovryn and H.W. Kraner, *J. Electrochem. Soc.*, 130 (1983) 917.
[35] C.G. Granqvist and I. Hamburg, *Proc. S.P.I.E.,* 562 (1985) 137.
[36] A. Corradini, A.M. Marinangeli and M. Mastragostino, *Electrochim. Acta*, 35 (1990) 1757.
[37] J.A. Duffy, M.D. Ingram and P.M.S. Monk, *Solid State Ionics,* 58 (1992) 109.
[38] M.D. Ingram, J.A. Duffy and P.M.S. Monk, *J. Electroanal. Chem.*, in press.
[39] S. Fletcher, L. Duff and R.G. Barradas, *J. Electroanal. Chem.*, 100 (1979) 759.
[40] M. Scholten and J.E.A.M. van der Meerakker, *J. Electrochem. Soc.*, 140 (1993) 471.
[41] P.M.S. Monk and C.M. Man: paper in preparation.
[42] S. Gottesfeld, J.D.E. McIntyre, G. Beni and J.L. Shay, *Appl. Phys. Lett.*, 33 (1978) 208.
[43] S. Gottesfeld and J.D.E. McIntyre, *J. Electrochem. Soc.*, 126 (1979) 742.
[44] G. Beni, C.E. Rice and J.L. Shay, *J. Electrochem. Soc.*, 127 (1980) 1342.
[45] K.S. Kang and J.L. Shay, *J. Electrochem. Soc.*, 130 (1983) 766.
[46] G. Beni and J.L. Shay, *Appl. Phys. Lett.*, 33 (1978) 567.
[47] J.L. Shay, G. Beni and L.M. Schiavone, *Appl. Phys. Lett.*, 33 (1978) 942.
[48] J.D.E. McIntyre, W.F. Peck Jr. and S. Nakahara, *J. Electrochem. Soc.*, 127 (1980) 1264.
[49] G. Beni and J.L. Shay, *Phys. Rev. B,* 21 (1980) 364.
[50] L.M. Schiavone, W.C. Dautremont-Smith, G. Beni and J.L. Shay, *Appl. Phys. Lett.*, 35 (1979) 823.

[51] W.C. Dautremont-Smith, G. Beni, L.M. Schiavone and J.L. Shay, *Appl. Phys. Lett.*, 35 (1979) 565.
[52] S. Gottesfeld and L.M. Schiavone, *J. Electroanal. Chem.*, 86 (1978) 89.
[53] C.E. Rice, *Appl. Phys. Lett.*, 35 (1979) 563.
[54] R. Sanjinés, A. Aruchamy and F. Lévy, *J. Electrochem. Soc.*, 136 (1989) 1740.
[55] S. Hackwood, L.M. Schiavone, W.C. Dautremont-Smith and G. Beni, Abstr. 511, p. 1262, *The Electrochemical Society,* extended abstracts, 1981, 80–81, quoted in reference [45] above.
[56] T. Yoshino, N. Baba and K. Arai, *Proc. Electrochem. Soc.*, 88 (1988) 23.
[57] K. Yamanaka, *Jpn. J. Appl. Phys.*, 30 (1991) 1285.
[58] C. Guterrez, M. Sanchez, J.I. Pena, C. Martinez and M.A. Martinez, *J. Electochem. Soc.*, 134 (1987) 2119.
[59] S. Ishihara *et al.*, *Jpn. Kokai Tokkyo Koho,* JP 63, 119,035, (1989), cited in *Chem. Abstr.* 110: P48,553*z*.
[60] J.L. Ord and D.J. De Smet, *J. Electrochem. Soc.*, 130 (1983) 280.
[61] S.K. Deb and J.A. Chopoorian, *J. Appl. Phys.*, 37 (1966) 4818.
[62] R.J. Colten, A.M. Guzman and J.W. Rabalais, *J. Appl. Phys.*, 49 (1978) 409.
[63] T.C. Arnoldssen, *J. Electrochem. Soc.*, 123 (1976) 527.
[64] A. Guerfi and L.H. Dao, *J. Electrochem. Soc.*, 136 (1989) 2435.
[65] P.M.S. Monk, T. Ali and R.D. Partridge, unpublished.
[66] B. Tell and F. Wudl, *J. Appl. Phys.*, 50 (1979) 5944.
[67] D.J. De Smet and J.L. Ord, *J. Electrochem. Soc.*, 134 (1987) 1734.
[68] S. Crouch-Baker and P.G. Dickens, *Mater. Res. Bull.*, 19 (1984) 1457.
[69] H. Suiyang, C. Fengbo and Z. Jicai, in B.V.R. Chowdari and S. Radhakrishna (eds.), 'Proceedings of the International Seminar on Solid State Ionic Devices', World Publishing Co., Singapore, 1988. p 521.
[70] W.C. Dautremont-Smith, *Displays*, 3 (1982) 3.
[71] J.N. Yao, B.H. Loo, K. Hashimoto and A. Fujishima, *Ber. Bunsenges. Phys. Chem.*, 95 (1991) 557.
[72] J.S.E.M. Svensson and C.G. Granqvist, *Appl. Phys. Lett.*, 49 (1986) 1566.
[73] A. Nemetz, A. Temmink, K. Bange, S. Cordoba de Torresi, C. Gabrielli, R. Torresi and A. Hugot le Goff, *Solar Energy Mater. Solar Cells*, 25 (1992) 93.
[74] R.C. Singh, S.S. Sekhon, K.S. Sidhu and S. Singh, *Extended Abstracts of the 7th International Conference on Solid State Ionics,* Japan, 1989, abstract 6pB–37.
[75] J. Scarminio, U. Urbano, B.J. Gardes and A. Gorenstein, *J. Mater. Sci. Lett.*, 11 (1992) 562.
[76] S. Passerini, B. Scrosati, A. Gorenstein and C.G. Granqvist, *J. Electrochem. Soc.*, 136 (1989) 3394.

[77] A.M. Andersson, C.G. Granqvist and J.R. Stevens, *Proc. S.P.I.E.*, 1016 (1988) 41.

[78] J.S.E.M. Svensson and C.G. Granqvist, *Appl. Optics*, 26 (1987) 1554.

[79] S. Yamada, T. Yoshioka, M. Miyashita, K. Urabe and M. Kikao, *Shizuoka Diagaku Denshi Kogaku Kenkyu Nokoku*, 22 (1987) 1, cited in *Chem. Abstr.* 107: 225,844*f*.

[80] C.G. Granqvist, A.M. Andersson and W. Estrade, *J. Appl. Phys.*, 64 (1988) 3678.

[81] M. Kitao and S. Yamada, in B.V.R. Chowdari and S. Radhakrishna (eds.), 'Proceedings of the International Seminar on Solid State Ionic Devices', World Publishing Co., Singapore, 1988, p 359.

[82] A.M. Andersson, W. Estrada and C.G. Granqvist, *Proc. S.P.I.E.*, 1272 (1990) 96.

[83] W. Estrada, A.M. Andersson, C.G. Granqvist, A. Gorenstein and F. Decker, *J. Mater. Res.*, 6 (1991) 1715.

[84] W. Visscher and E. Barendrecht, *J. Electroanal. Chem.*, 154 (1983) 69.

[85] M.K. Carpenter, R.S. Connell and D.A. Corrigan, *Solar Energy Mater.*, 16 (1987) 333.

[86] R.M. Bendert and D.A. Corrigan, *J. Electrochem. Soc.*, 136 (1989) 1369.

[87] R.M. Bendert and D.A. Corrigan, *J. Electrochem. Soc.*, 136 (1989) 723.

[88] C. Liquan, D. Ming, C. Yunfa, S. Chunxiang and X. Rungjian, *Extended Abstracts of the 7th International Conference on Solid State Ionics*, Japan, 1989, abstract 6pB–38.

[89] P.C. Yu, G. Nazri and C.M. Lampert, *Solar Energy Mater.*, 16 (1987) 1.

[90] M. Fantini and A. Gorenstein, *Solar Energy Mater.*, 16 (1987) 487.

[91] R.S. Conell, D.A. Corrigan and B.R. Powell, *Solar Energy Mater. Solar Cells*, 25 (1992) 307.

[92] R.M. Torresi, M.V. Vásquez, A. Gorenstein and S.I. Córdoba-Torresi, *Thin Solid Films*, 1993

[93] A. Hugot-Le Goff and S. Cordoba de Torresi, *Proc. Electrochem. Soc.*, 90–2 (1990) 157.

[94] D.A. Corrigan and S.L. Knight, *J. Electrochem. Soc.*, 136 (1989) 613.

[95] F. Decker, S. Passerini, R. Pileggi and B. Scrosati, *Electrochim. Acta*, 37 (1992) 1033.

[96] J.L. Ord, *Surf. Sci.*, 56 (1976) 413.

[97] A. Gorenstein, F. Decker, W. Estrade, C. Esteres, A. Andersson, S. Passerini, S. Pantaloni and B. Scrosati, *J. Electroanal. Chem.*, 277 (1990) 277.

[98] S.K. Deb, *Appl. Optics*, supp. 3 (1969) 192.

[99] B.W. Faughnan and R.S. Crandall, in J.I. Pankove (ed.), 'Display Devices', Springer-Verlag, Berlin, 1980. Chapter 5.

[100] G.V Granqvist, *Phys. Thin Films*, 17 (1993) 301.

[101] R.B. Goldner, in B.V.R. Chowdari and S. Radhakrishna (eds.), 'Proceedings of the International Seminar on Solid State Ionic Devices, World Publishing Co., Singapore, 1988. p 379.
[102] R.D. Varjian, M. Shabrand and S. Babinac, *Proc. Electrochem. Soc.*, 94–2 (1994) 278.
[103] N. Kaneto, J. Tobata and T. Miyoshi, *Society of Information Displays International Symposium,* 12 (1981) 74.
[104] F.G.K. Baucke, *Schott Information,* 1 (1983) 11.
[105] F.G.K. Baucke and J.A. Duffy, *Chem. Brit.,* 21 (1985) 643.
[106] F.G.K. Baucke, *Rivista della Staz. Sper. Vetro,* 6 (1986) 119.
[107] F.G.K. Baucke, *Feinwerktechnik und Meßtechnik,* 94 (1986) 25.
[108] F.G.K. Baucke, *Solar Energy Mater.,* 16 (1987) 67.
[109] K. Matsuhiro and Y. Masuda, *Proc. S.I.D.,* 21 (1980) 101.
[110] H. Witzke and S.E. Schalterly, (Optic), US pat: 3,840,287 (1974).
[111] H.N. Hersch, W.E. Kramer and J.H. McGee, *Appl. Phys. Lett.,* 27 (1975) 646.
[112] S.K. Mohapatra, G.D. Boyd, F.G. Storz, S. Wagner and F. Wudl, *J. Electrochem. Soc.,* 126 (1979) 805.
[113] A.T. Howe, S.H. Sheffield, P.E. Childs and M.G. Shilton, *Thin Solid Films,* 67 (1980) 365.
[114] R.D. Giglia and G. Haacke, *Society of Information Displays International Symposium,* 12 (1981) 76.
[115] C. Cohen, *Electronics,* 11 (1981) 65.
[116] P. Schlotter and L. Pickelmann *J. Electron. Mater.,* 11 (1982) 207.
[117] T. Kamimori, J. Nagai and M. Mizuhashi, *Solar Energy Mater.,* 16 (1987) 27.
[118] B.W. Faughnan, R.S. Crandall and P. M. Heyman, *R.C.A. Rev.,* 36 (1975) 177.
[119] J.-P. Randin, *J. Electronic Mater.,* 7 (1978) 47.
[120] R. Hurditch, *Electron. Lett.,* 11 (1975) 142.
[120a] S.J. Babinec, *Solar Energy Mater. Solar Cells,* 25 (1992) 269.
[121] R.J. Stocker, S. Singh, L.G. van Uitert and G.J. Zydzik, *J. Appl. Phys.,* 50 (1979) 2993.
[122] O. Bohnke, C. Bohnke, G. Robert and B. Carquille, *Solid State Ionics,* 6 (1982) 121.
[123] J.A. Duffy, M.D. Ingram and P.M.S. Monk, *Solid State Ionics,* 58 (1992) 109.
[124] B. Reichman and A.J. Bard, *J. Electrochem. Soc.,* 126 (1979) 583.
[125] M.M. Hefny, *Bull. Electrochem.,* 3 (1987) 11.
[126] B. Scrosati, in B.V.R. Chowdari and S. Radhakrishna (eds.), 'Proceedings of the International Seminar on Solid State Ionic Devices', World Publishing Co., Singapore, 1988. p 341.
[127] P. Pedone, M. Armand and D. Deroo, *Solid State Ionics,* 28–30 (1988) 1729.
[128] B. Tell, *J. Electrochem. Soc.,* 127 (1980) 2451.

[129] P.K. Shen, H. Huang and A.C.C. Tseung, *J. Mater. Chem.*, 2 (1992) 497.

[130] M. Green, D.C. Smith and J.A. Weiner, *Thin Solid Films*, 38 (1976) 89.

[131] L.Y. Berezin and V.P. Malinenko, *Pis'ma. Zh. Tekh. Fiz.*, 13 (1987) 401, cited in *Chem. Abstr.* 107: 49,382*t*.

[132] S.K. Deb, *Phil. Mag.*, 27 (1973) 801.

[133] M.S. Whittingham, in B.V.R. Chowdari and S. Radhakrishna (eds.), 'Proceedings of the International Seminar on Solid State Ionic Devices', World Publishing Co., Singapore, 1988. p 325.

[134] J.S.E.M. Svensson and C.G. Granqvist, *Appl. Phys. Lett.*, 45 (1984) 828.

[135] R.B. Goldner, P. Norton, K. Wong, G. Foley, E.L. Goldner, G. Seward and R. Chapman, *Appl. Phys. Lett.*, 47 (1985) 536.

[136] G.G. Barna, *J. Electron. Mater.*, 8 (1979) 153.

[137] O. Bohnke, C. Bohnke, A. Donnadieu and D. Davazoglou, *J. Appl. Electrochem.*, 18 (1988) 447.

[138] A. Donnadieu, D. Davazoglou and A. Abdellaoui, *Thin Solid Films*, 164 (1988) 333.

[139] D. Davazoglou, A. Donnadieu and O. Bohnke, *Solar Energy Mater.*, 16 (1987) 55.

[140] A. Donnadieu, M. Regrazini, B. Ducourant, R. Foucade and O. Bohnke, *Proc. S.P.I.E.*, 1728 (1992) 2.

[141] A. Deneuville, P. Gérard and R. Billat, *Thin Solid Films*, 70 (1980) 203.

[142] J. Livage, A. Zarudiansky, R. Rose and P. Judenstein, *Solid State Ionics*, 28–30 (1988) 1722.

[143] P. Judinstein and J. Livage, *Proc. S.P.I.E.*, 1328 (1990) 344.

[144] P. Judinstein and J. Livage, *J. Mater. Chem.*, 1 (1991) 621.

[145] M.I. Yarovskaya, I. E. Obvintseva, V.G. Kessler, B. Sh. Galyamov, S.I. Kucheiko, R.R. Shifrina and N.Y. Turova, *J. Non-Cryst. Solids*, 124 (1990) 155.

[146] H. Dislich, *Le Vide-Les Couches Minces*, 227 (1985) 261.

[147] J.A. Duffy and R.I. Smith: unpublished.

[148] M.T. Nguyen and L.H. Dao, *Proc. Electrochem. Soc.*, 90–2 (1990) 246.

[149] P.K. Shen, J. Syed-Bokhari and A.C.C. Tseung, *J. Electrochem. Soc.*, 138 (1991) 2778.

[150] P.M.S. Monk and S.L. Chester, *Electrochim. Acta*, 38 (1993) 1521.

[151] Y. Shimizu, K. Noda, K. Nagase, N. Miura and N. Yamazoe, *Sogo Rikogaku Kenkyuka Hokoku (Kyusha Daigaku Diagakuin)*, 12 (1991) 367, cited in *Chem. Abstr.* 115: 102,676*k*.

[152] K. Yamanaka, H. Oakamoto, H. Kidoi and T. Kudo, *Jpn. J. Appl. Phys.*, 25 (1986) 1420.

[153] S.F. Cogan, T.D. Plante, M.A. Parker and R.D. Rauh, *J. Appl. Phys.*, 60 (1986) 2735.

[154] B.W. Faughnan, R.S. Crandall and M.A. Lampert, *Appl. Phys. Lett.*, 27 (1975) 275.

[155] P.G. Dickens, D.J. Murphy and T.K. Holstead, *J. Solid State Chem.*, 6 (1973) 370.

[156] M.A. Vanice, M. Boudart and J. J. Fripiat, *J. Catalysis,* 17 (1970) 359.

[157] S. Kurita, T. Nishimura and K. Taira, *Appl. Phys. Lett.*, 36 (1980) 585.

[158] C.-K. Ho, I.D. Raistrick and R.A. Huggins, *J. Electrochem. Soc.*, 127 (1980) 343.

[159] C. Bohnke and O. Bohnke, *J. Appl. Electrochem.*, 18 (1988) 715.

[160] O. Bohnke and B. Vuillemin, in M. Balkanski, T. Takahashi and H.L. Tuller (eds.), 'Solid State Ionics', Elsevier, 1992. p 593.

[161] N. Yoshiike and S. Kondo, *J. Electrochem. Soc.*, 131 (1984) 809.

[162] S.K. Mohapatra, *J. Electrochem. Soc.*, 125 (1978) 284.

[163] O. Bohnke, thése d'etat, Besancon, France, 1984.

[164] C. Bohnke, thése d'etat, Besancon, France, 1986.

[165] P. Baudry, M.A. Aegerter, D. Deroo and B. Valla, *Proc. Electrochem. Soc.*, 90–2 (1990) 274.

[166] M. Green, *Thin Solid Films*, 50 (1978) 148.

[167] G. Xu and L. Chen, *Solid State Ionics*, 28–30 (1988) 1726.

[168] B.W. Faughnan and R.S. Crandall, *Appl. Phys. Lett.*, 26 (1975) 120.

[169] J.-P. Randin and P. Viennet, *J. Electrochem. Soc.*, 129 (1982) 2349.

[170] B. Reichman, A.J. Bard and D. Lasar, *J. Electrochem. Soc.*, 127 (1980) 647.

[171] M.S. Whittingham and R.A. Huggins, in W. van Gool (ed.), 'Fast Ion Transport Through Solids', North Holland, 1973.

[172] R.B. Goldner, in B.V.R. Chowdari and S. Radhakrishna (eds.), 'Proceedings of the International Seminar on Solid State Ionic Devices', World Publishing Co., Singapore, 1988. p 351.

[173] S. Joo, I.D. Raistrick and R.A. Huggins, *Solid State Ionics,* 17 (1985) 313.

[174] P. Gerard, A. Deneuville, G. Hollinger and T.M. Duc, *J. Appl. Phys.*, 48 (1977) 4252.

[175] L. Zhang and K.S. Goto, *Proc. Electrochem. Soc.*, 90–2 (1990) 23.

[176] P.G. Dickens and M.S. Whittingham, *Quart. Rev. Chem. Soc.*, 22 (1968) 30.

[177] O.F. Schirmer, V. Wittner, G. Baur and G. Brandt, *J. Electrochem. Soc.*, 124 (1977) 749.

[178] J.F. Owen, K.J. Teegarden and H.R. Shanks, *Phys. Rev. B*, 18 (1978) 3827.

[179] A. Deneuville and P. Gérard, *J. Electron. Mater.*, 7 (1978) 559.

[180] W.C. Dautremont-Smith, M. Green and K.S. Kang, *Electrochim. Acta,* 22 (1977) 751.

[181] R.S. Crandall and B.W. Faughnan, *Phys. Rev. Lett.*, 39 (1977) 232.

[182] J.B. Goodenough, *Prog. Solid. Stat. Chem.*, 5 (1971) 315.

[183] R.B. Goldner, D.H. Mendelsohn, J. Alexander, W.R. Henderson, D. Fitzpatrick, T.E. Haas, H.H. Sample, R.D. Rauh, M.A. Parker and T.L. Rose, *Appl. Phys. Lett.*, 43 (1983) 1093.

[184] R.B. Goldner and D.H. Mendelsohn, *J. Electrochem. Soc.*, 131 (1984) 857.

[185] R.B. Goldner, P. Norton, K. Wong, G. Foley, E.L. Goldner, G. Seward, and R. Chapman, *Appl. Phys. Lett.*, 47 (1985) 536.

[186] V. Wittner, O.F. Schirmer, and P. Schlotter, *Solid State Commun.*, 25 (1978) 977.

[187] I.F. Chang, B.L. Gilbert and T.I. Sun, *J. Electrochem. Soc.*, 122 (1975) 955.

[188] Y.S. Krasnov, O.A. Sych and F.N. Patsyuk, *Elektrokhimiya,* 24 (1988) 1468.

[189] N.F. Mott, 'Conduction in Non-Crystalline Materials', 2nd Edn., Clarendon Press, Oxford, 1993.

[190] P.A. Cox, 'The Electronic Structure and Chemistry of Solids', Oxford University Press, Oxford, 1987

[191] P.A. Cox, 'Transition Metal Oxides: An Introduction to their Electronic Structure and Properties', International Series of Monographs in Chemistry No. 27, Clarendon Press, Oxford, 1992.

[192] J.H. Pfifer and E.K. Sichel, *J. Electron. Mater.*, 9 (1980) 129.

[193] J.A. Duffy, F.G.K. Baucke and R.I. Smith, *Thin Solid Films,* 186 (1990) 47.

[193a] F.G.K. Baucke, K. Bange and T. Gampke, *Displays,* (1988) 179.

[194] M. Green, W.C. Dautremont-Smith and K.S. Kang, *2nd International Conference on Solid Electrolytes (St. Andrews, Scotland, UK),* 1978. (reference [30] of our reference [67] above).

[195] S. Yamada, S. Yoshida and M. Kikao, *Extended Abstracts of the 7th International Conference on Solid State Ionics,* Japan., 1989, abstract 6pB–34.

[196] M.S. Burdis and J.R. Siddle, *Thin Solid Films*, 237 (1994) 320.

[197] J.L. Ord, *J. Electrochem. Soc.*, 129 (1982) 767.

[198] J.L. Ord, G.M. Pepin and D.J. Beckstead, *J. Electrochem. Soc.*, 136 (1989) 362.

[199] Y. Fujita, K. Miyazaki and C. Tatsuyama, *Jpn. J. Appl. Phys.*, 24 (1985) 1082.

[200] S.F. Cogan, N.M. Nguyen, S.J. Perrotti and R.D. Rauh, *J. Appl. Phys.*, 66 (1989) 1333.

[201] S.F. Cogan and R.D. Rauh, *Solid State Ionics,* 28–30 (1988) 1707.

[202] S.D. Hansen and C.R. Aita, *J. Vac. Sci. Technol. (A),* 3 (1985) 660.

[203] S.F. Cogan, R.D. Rauh, J.D. Westwood, T.D. Plante and N.M. Nguyen, *Proc. Electrochem. Soc.*, 90–2 (1990) 99.

[204] S.F. Cogan, N.M. Nguyen, S.J. Perrotti and R.D. Rauh, *Proc. S.P.I.E.,* 1016 (1988) 57.

[205] C.R. Aita, Y. Liu, M.L. Kao and S.D. Hansen, *J. Appl. Phys.,* 60 (1986) 749.

[206] Y. Shimizu, K. Nagase, N. Muira and N. Yamazoe, *Solid State Ionics,* 53–56 (1992) 490.

[207] Y. Shimizu, K. Nagase, N. Muira and N. Yamazoe, *Appl. Phys. Lett.*, 60 (1992) 802.

[208] A. Talledo, A.M. Andersson and C.G. Granqvist, *J. Appl. Phys.*, 69 (1991) 3261.

[209] A.I. Gavrilyuk and F.A. Chudnovski, *Sov. Tech. Phys. Lett.*, 3 (1977) 69.

[210] S. Hub, A. Trenchant and R. Messina, *Electrochim. Acta*, 33 (1988) 997.

[211] D.W. Murphy and P.A. Christian, *Science*, 205 (1979) 651.

[212] P.G. Dickens and G.J. Reynolds, *Solid State Ionics*, 5 (1981) 331.

[213] H.G. Bachmann, F.R. Ahmed and W.H. Barnes, *Z. Kristall. Bd*, 115 (1961) 110.

[214] D.W. Murphy, P.A. Christian, R.J. Disalvo and J.V. Waszczak, *Inorg. Chem.*, 18 (1979) 288.

[215] J.C. Clayton and D.J. De Smet, *J. Electrochem. Soc.*, 134 (1987) 1734.

[216] J.L. Ord, S.D. Bishop and D.J. De Smet, *Proc. Electrochem. Soc.*, 90–2 (1990) 99.

[217] J. Scarminio, A. Talledo, A.A. Andersson, S. Passerini and F. Decker, *Electrochim. Acta*, 38 (1993) 1637.

[218] M.S.R. Khan, K.A. Khan, W. Estrada and C.G. Granqvist, *J. Appl. Phys.*, 69 (1991) 3231.

[219] P. Baudry, A.C.M. Rodriguez, M.A. Aegerter and L.O. Bulhôes, *J. Non-Cryst. Solids*, 121 (1990) 319.

[220] U. Stangar, B. Orel, I. Grabec, B. Ogorevc and K. Kalcher, *Solar Energy Mater. Solar Cells*, 31 (1993) 171.

[221] P. Baudry, Ph.D thesis, 1990, Grenoble, France.

[222] L.D. Burke and O.J. Murphy, *J. Electroanal. Chem.*, 109 (1980) 379.

[223] C. Gutierrez and B. Beden, *J. Electroanal. Chem.*, 293 (1990) 253.

[224] L.D. Burke and O.J. Murphy, *J. Electroanal. Chem.*, 109 (1980) 373.

[225] S.I. Córdoba de Torresi and A. Gorenstein, *Electrochim. Acta*, 37 (1992) 2015.

[226] N. Baba, T. Yoshino and S. Watanabe, *Extended Abstracts of the 7th International Conference on Solid State Ionics, Japan*, 1989, abstact 6pB–39.

[227] M.C. Bernard, S. Cordoba de Torresi, A. Hugot-Le Goff and V.T. Bich, *J. Electrochem. Soc.*, 140 (1993) 3065.

[228] M.C. Bernard, S. Cordoba de Torresi, A. Hugot-Le Goff and B. Vu Thi, *Proc. Electrochem. Soc.*, 94–2 (1994) 146.

[229] F. Garnich, P.C. Yu and C.M. Lampert, *Solar Energy. Mater.*, 20 (1990) 265.

[230] O. Erlandsson, J. Lindvall, N.T. Nguyen, V.H. Nguyen, V.T. Bich and N.D. Nguyen, *Proc. S.P.I.E.*, 2017 (1993) 192.

[231] O. Erlandsson, J. Lindvall, N.T. Nguyen, N.N. Toan, N.V. Hung V.T. Bich and N.N. Dinh, *Phys. Status Solidi A*, 139 (1993) 451.

[232] Y.I. Malnyk and L.I. Skathov, *Pis'ma. Zh. Tekh. Fiz.*, 13 (1987) 504, cited in *Chem. Abstr.* 107: 207,631*x*.

[233] S.F. Cogan, T.D. Plante, M.A. Parker and R.D. Rauh, *Solar Energy Mater.*, 14 (1986) 185.

[234] S.F. Cogan, T.D. Plante, E.J. Anderson and R.D. Rauh, *Proc. S.P.I.E.*, 562 (1985) 23.

[235] R. Cabanel, J. Chaussy, J. Mazuer, G. Delabouglise, J.C. Joubert, G. Barral and C. Montella, *J. Electrochem. Soc.*, 137 (1990) 1444.

[236] M.A.B. Gomes, L.O. Bulhôes, S.C. De Castro and A.J. Damiao, *J. Electrochem. Soc.*, 137 (1990) 3067.

[237] G.R. Lee and J.A. Crayston, *J. Mater. Chem.*, 1 (1991) 381.

[238] A.Z. Bolzan and A.J. Arvia, *J. Electroanal. Chem.*, 322 (1992) 247.

[239] L.D. Burke and E.J.M. O'Sullivan, *J. Electroanal. Chem.*, 93 (1978) 11.

[240] S. Gottesfeld, *J. Electrochem. Soc.*, 127 (1980) 272.

[241] L.D. Burke and D.P. Whelan, *J. Electroanal. Chem.*, 103 (1979) 179.

[242] E. Inoue, H. Kokado and A. Izawa, *Oyo Butsusi*, 43 (1974) 54.

[243] T. Ohzuki and T. Hirai, *Electrochim. Acta*, 27 (1982) 1263

[244] M. Ottaviani, S. Panero, S. Morizilli, B. Scrosati and M. Lazzari, *Solid State Ionics*, 20 (1986) 197.

[245] J.L. Ord, D.J. De Smet and D.J. Beckstead, *J. Electrochem. Soc.*, 136 (1989) 2178.

[246] K.K. Lian and V.I. Birss, *J. Electrochem. Soc.*, 138 (1991) 2877.

[247] P.M.S. Monk, S.L. Chester, D.S. Higham and R.D. Partridge, *Electrochim. Acta.*, 39 (1994) 2277.

[248] B.W. Faughnan and R.S. Crandall, *Appl. Phys. Lett.*, 31 (1977) 834.

[249] D.A. Corrigan and M.K. Carpenter, *S.P.I.E. Institute Series*, IS4 (1990) 298

[250] D.A. Corrigan, *Solar Energy Mater. Solar Cells*, 25 (1992) 293.

[251] Y.P. Ma, P.C. Yu and C.M. Lampert, *Proc. S.P.I.E.*, 1536 (1991) 93.

[252] E.K. Sichel and G.I. Gittleman, *J. Electronic Mater.*, 8 (1979) 1.

[253] B.J. Molnar, A.R. Haranahalli and B.D. Dove, *J. Vac. Sci. Technol.*, 15 (1978) 161.

[254] P.K. Shen and A.C.C. Tseung, *J. Mater. Chem.*, 2 (1992) 1141.

[255] L.H. Dao, A. Guerfi and M.T. Nguyen, *Proc. Electrochem. Soc.*, 90–2 (1990) 30.

[256] P.V. Ashrit, G. Bader, F.E. Giround and V.-V. Truong, *Proc. Electrochem. Soc.*, 90–2 (1990) 45.

[257] H. Matsuoka, S. Hashimoto and H. Kagechika, *Hyomen Gijutsu*, 42 (1991) 246 cited in *Chem. Abstr.* 115: 243,788*k*.

[258] S. Hashimoto and H. Kagechika, *J. Electrochem. Soc.*, 138 (1991) 2403.

[259] J. Gottsche, A. Hinsch and V. Wittwer, *Solar Energy Mater. Solar Cells*, 31 (1993) 415.

[260] S. Hashimoto and H. Matsuoka, *Surface and Interface Analysis*, 19 (1990) 464; *J. Electrochem. Soc.*, 138 (1990) 2403.

[261] P.A. Gillet, J.L. Fourquet and O. Bohnke, *Mat. Res. Bull.*, 27 (1992) 1145.

[262] P.A. Gillet, J.L. Fourquet and O. Bohnke, *Proc. S.P.I.E.*, 1728 (1992) 82.

[263] K. Nagase, Y. Shimizu, N. Miura and N. Yamazoe, Abstr. 1484, *Extended Abstracts of the 183rd Electrochemistry Society Meeting*, Hawaii, 1993.

[264] S.F. Cogan, R.D. Rauh, N.M. Nguyen, T.D. Plante and J.D. Westwood, *J. Electrochem. Soc.*, 140 (1993) 112.

[265] K. Nagase, Y. Shimizu, N. Miura and N. Yamazoe, *J. Ceram. Soc. Jpn.*, 101 (1993) 1032.

[266] K. Nagase, Y. Shimizu, N. Miura and N. Yamazoe, *App. Phys. Lett.*, 61 (1992) 243.

[267] B. Valla, J.C.L. Tonazzi, M.A. Macedo, L.H. Dall'Antonio, M.A. Aegerter, M.A.B. Gomes and L.O. Bulhôes, *Proc. S.P.I.E.*, 1536 (1991) 48

[268] P. Baudry, A.C.M. Rodrigues, M.A. Aegerter and L.O. Bulhôes, *J. Non-Cryst. Solids*, 121 (1990) 319.

[269] M. Vukovic, D. Cukman, M. Milun, L.D. Atanasoska and R.T. Atanasoska, *J. Electroanal. Chem.*, 330 (1992) 663.

[270] R. Braunstein, *Solid State Commun.*, 28 (1978) 839.

[271] W.F. Chu, R. Hartman, V. Leonhard and G. Ganson, *Mat. Sci. and Eng.*, B13 (1992) 235

[272] C. Ruf, K. Bärmer and R. Braunstein, *Solid State Commun.*, 54 (1985) 111.

[273] P.M.S. Monk, R.D. Partridge, R. Janes and M. Parker, *J. Mater. Chem.*, 4 (1994) 1071.

[274] K. Hashemi, J. Illingworth and A.W. Brinkman, *J. Mater. Sci. Lett.*, 9 (1990) 1176.

[275] J. Illingworth, K. Hashemi and A.W. Brinkman, Applied Solid-State Chemistry Group (RSC) symposium 'New Applications of Electrochromism: Display, Light Modulation and Printing' Meeting, London , 3rd April 1991.

[276] H. Yoneyama and Y. Shoji, *J. Electrochem. Soc.*, 137 (1990) 3826.

[277] M. Dahlhaus and F. Beck, *J. Appl. Electrochem.*, 23 (1993) 957.

[278] P.K. Shen, H.T. Huang and A.C.C. Tseung, *J. Electrochem. Soc.*, 139 (1992) 1840.

[279] H. Yoneyama, S. Hirao and S. Kuwabata, *J. Electrochem. Soc.*, 139 (1992) 3141.

[280] M. Morita, *Macromol. Chem. Phys.*, 195 (1994) 609.

5 Phthalocyanine Compounds

5.1 Introduction

Molecular metallo-organic phthalocyanines have been employed as pigments in the dyestuffs industry for many years, but recently new rare-earth phthalocyanines have been used as electrochromic species in ECDs.

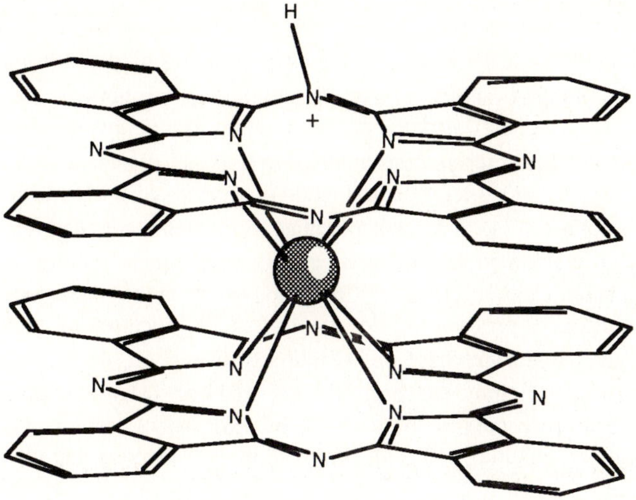

Fig. 5.1 Structure of lutetium bis(phthalocyanine).

The phthalocyanine ring is part of the structure shown in Fig. 5.1. Metallo-organic complexes may take two stoichiometries, either with a metal ion residing at the centre of a single phthalocyanine ring or, for the more common *bis*(phthalocyanines), between two rings in a sandwich-type compound. Reduction occurs at the rings; electron uptake by the Lu can cause molecular dissociation.

The rare-earth phthalocyanines are generally prepared by the method of Moskalev and Kirin [1] in which a rare-earth metal acetate reacts with 1,2-dicyanobenzene. Samples made in this way are best purified by sublimation, forming thin films of rare-earth phthalocyanine. Such films are vividly coloured even in their neutral form.

5.2 Lutetium *bis*(Phthalocyanine)

The phthalocyanine compound which has received the most attention is lutetium *bis*(phthalocyanine), $Lu(pc)_2$, where 'pc' represents one phthalocyanine ring. $Lu(pc)_2$ has been studied extensively by Collins and Schiffrin [2, 3] and by Nicholson and co-workers [4–12]. Collins and Schiffrin's $Lu(pc)_2$ was initially studied as a film immersed in aqueous electrolyte but such solvents were undesirable, however, as hydroxide ion from water caused gradual film destruction, attacking nitrogens of the pc ring [3].

Acidic solution allows a greater number of write–erase cycles, for example 5×10^6 write–erase cycles in sulphuric acid [3] are quoted. $Lu(pc)_2$ films in ethylene glycol solution [2] were subsequently studied and found to be 'at least three orders of magnitude' more stable. Nicholson studied solid phthalocyanine films in aqueous electrolytes [5, 10], and soluble species in organic solution [5, 10].

Fresh $Lu(pc)_2$ films are brilliant green in colour (λ_{max} = 605 nm: see table 5.1). There is much evidence to show this form of $Lu(pc)_2$ to be singly protonated [12] as $[pc–Lu–pc–H]^+$; the phthalocyanines of uranium and thorium are not electrochromic unless protonated [14]. The green $Lu(pc)_2$ may be oxidised to a yellow/tan form [5, 10, 15]. Chang and Marchon [15] prepared this oxidised species as a diamagnetic salt by chemical oxidation of green $Lu(pc)_2$. A further oxidation product is red [5, 10, 15]. Alternatively, electroreduction of green lutetium phthalocyanine forms a blue-coloured film [16, 17], and further reduction yields a violet/blue product [10]. Agreement between various groups concerning film composition is tenuous: Chang and Marchon [15] doubt the occurrence of the above-mentioned protonation in green $Lu(pc)_2$ because of data from mass spectroscopy, and Collins and Schiffrin [2] similarly dismiss it.

Electrochromic switching has been studied by chemical reduction coupled with magnetic susceptibility measurements [15], by ESR spectroscopy [15, 18] and by radioactive isotopic labelling [10, 14]. While study has concentrated on solid $Lu(pc)_2$ films [5, 10, 14, 16, 18], there has also been work on phthalocyanine species electro-generated in solution [15, 18]. The preferred solvent for solution-phase electrochemistry is DMF [15]. The colours obtained for lutetium phthalocyanine as a thin film are summarised in table 5.1, together with spectroscopic data and proposed compositions. In table 5.2, data are presented concerning lutetium *bis*(phthalocyanine) in solution and in various different oxidation states.

In summary: the red, yellow (or brown) products involve the loss of 1 or 2 electrons from the green form of $Lu(pc)_2$, and reduction to the blue and violet forms result from the uptake of 1, 2 or 3 electrons [12]. The electrochemistry of oxidation and reduction of the $Lu(pc)_2$ films is discussed in a short but informative review by Nicholson [12].

Plichon *et al.* [19] have used the mirage effect – deflection of a laser beam during passage through a solution layer of variable refractive index – to identify which redox processes involve anion and which cation movement into thin-film $Lu(pc)_2$.

The lutetium *bis*(phthalocyanine) system is a truly polyelectrochromic one, and has been recognised as such since 1970 [20], but usually only the blue-to-green transition is used in a bi-electrochromic device. Although many prototypes have been constructed [4, 21, 22], no ECD incorporating $Lu(pc)_2$ has yet been marketed, owing to experimental difficulties such as film disintegration caused by constant anion insertion and egress

Table 5.1 Colours, Wavelength Maxima and Suggested Composition of Lutetium *bis*(phthalocyanine) Redox States as Solid Films.

Colour	λ_{max}/nm	Proposed Formula	ref.
green	605	$[pc\ Lu\ pc\ H]^+$	[2]
Anodic Products			
yellow	682	$[pc\ Lu\ pc\ H]^+\ Cl^-$	[12]
red	495, 695	$[pc\ Lu\ pc\ H]^{2+}\ 2Cl^-$	[5, 10]
Cathodic Products			
light blue	—	$[pc\ Lu\ pc\ H]^-\ Li^+$	[10]
blue	—	$[pc\ Lu\ pc\ H]^{2-}\ 2K^+$	[10]
dark-blue/violet	—	$[pc\ Lu\ pc\ H_{n+1}]$ $2 \leq n \leq 4$	[10]

Table 5.2 Colours, Wavelength Maxima and Suggested Composition of Lutetium *bis*(phthalocyanine) Redox States in Solution.

Colour	Solvent	λ_{max}/nm	Magn.[a]	Proposed Formula	ref.
green	DMF	600	para.	$[pc\ Lu\ pc\ H]^+$	[18]
Anodic Products					
yellow-tan	CH_2Cl_2 chem. redn.	474, 688	dia.	$[pc\ Lu\ pc\ H]^+\ SbCl_6^-$	[15]
yellow-red	DMF	690	dia.		[18]
red	DMF	—	—		[18]
Cathodic Products					
blue	DMF	612	dia.	$[pc\ Lu\ pc\ H]$	[18]
violet	DMF	—	para.	$[pc\ Lu\ pc\ H]^-$	[18]
violet	CH_2Cl_2 chem. redn.	341, 520 620	para.	$[pc\ Lu\ pc\ H]^{2-}$	[15]

[a] Magnetic behaviour: 'para.' = paramagnetic and 'dia.' = diamagnetic.

during electro-switching [2]. For this reason, larger anions are best avoided to minimise the mechanical stresses.

A second, related, handicap of phthalocyanine electrochromic devices is their relatively slow response times. Nicholson investigated the kinetics of colour reversal [11]: small anions like chloride and bromide allow faster colour switching. The kinetics of anion uptake have been studied by radio-tracer techniques [11] and by a novel moving colour-boundary experiment [5]. Castaneda *et al.* [24] also found ECD devices made with phthalocyanines to be slow at colour switching. The response time of $Lu(pc)_2$ films in solid polyethylene oxide (PEO) electrolytes is again found to be poor [23].

Sammells *et al.* [21, 22] have sidestepped the problem of the slow penetration of anions into solid lattices by using an ECD containing an electrochrome suspension in semi-solid polyAMPS electrolyte.

While the response times are somewhat slow, the open-circuit life-times ('memory') of all colours were found to be very good [11]. Films in chloride, bromide, iodide and sulphate-containing solutions were found to be especially stable in this respect.

5.3 Other Metal Phthalocyanines

Of the metal phthalocyanines, lutetium *bis*(phthalocyanine) has received overwhelmingly the most attention, although many other phthalocyanines have been prepared and studied. Moskalev *et al.* [16] prepared the phthalocyanines of neodymium, americium, europium, thorium and gallium (the latter as the half acetate). Collins and Schiffrin [2] have reported the electrochromic nature of the phthalocyanines of cobalt, tin(IV), tin(IV) dichloride, molybdenum and copper.

The same authors have also studied metal free (that is, di-hydrogen) phthalocyanine [2]. No electrochromic behaviour was observed for either the metal-free or for the copper phthalocyanines in the potential ranges employed. All the other salts made by Collins and Schiffrin showed (limited) electrochromism. Both tin phthalocyanines could be readily reduced, but showed no anodic electrochromism.

Other molecular phthalocyanine electrochromes studied include aluminium [25], copper [26], chromium [25, 27], erbium [28, 29], europium [30], iron [31], magnesium [32], manganese [32, 27], thorium [14], titanium [33], uranium [14, 34], vanadium [33], ytterbium [35, 36], zinc [33, 37] and zirconium [30, 38, 39], much of this work being performed by Silver and co-workers. This same group have also prepared a mixed phthalocyanine system by reacting mixed-metal precursors comprising the rare-earth metals dysprosium, holmium, erbium, thulium, ytterbium, lutetium, yttrium and small amounts of others [40]. The response times τ for such mixtures is reportedly superior to that for single-component films.

Walton *et al.* [41] have compared the electrochemistry of lutetium and ytterbium *bis*(phthalocyanines), finding them to be essentially identical, although $Yb(pc)_2$ did not exhibit an ESR signal in any of its redox states *cf.* table 5.2.

Transition-metal phthalocyanines contain only a single pc ring while lanthanide-containing species form *bis*(phthalocyanines). Both chromium and manganese mono-phthalocyanines undergo valence change during oxidation or reduction [27]. By contrast electrochemistry of $Lu(pc)_2$ occurs generally on the ligand; electron transfer to the central lutetium causes molecular dissociation [42].

Lever and co-workers have studied cobalt phthalocyanine systems in which two or four Co(pc) units are connected via chemical links [43–46]. These workers have also studied tetrasulphonated cobalt and iron phthalocyanines [47].

Finally, polymeric ytterbium *bis*(phthalocyanine) has been studied by Kashiwazaki and co-workers [48–52] using a plasma technique to effect the polymerisation.

5.4 Related Species

Cobalt octamethoxy-phthalocyanine has been studied by Collins and Schiffrin [2], as have octacyanophthalocyanines [53]. Lutetium *bis*(octaalkylphthalocyanine) has also been reported [54].

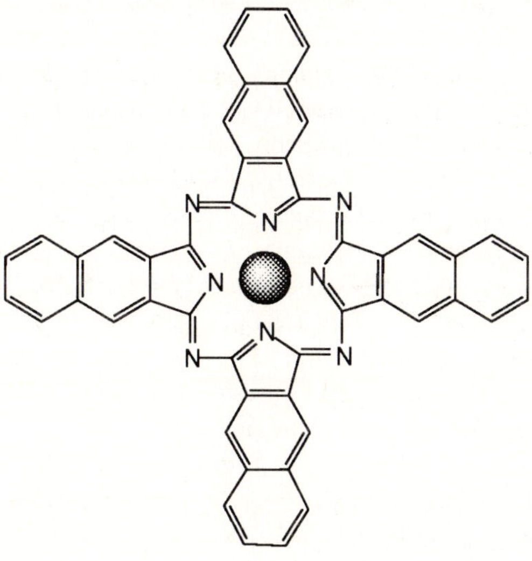

Fig. 5.2 Structure of naphthalocyanine (naph) showing metal ion residing at the central cavity of the ring.

Naphthalocyanine (naph) species (Fig. 5.2) are structurally similar to the simpler phthalocynanines above. Such species show an intense optical absorption at high wavelengths ($700 < \lambda < 900$ nm) [55–58] owing to electronic processes within the extended conjugation system of the ligand [56, 58].

Thin-film cobalt *bis*(naphthalocyanine) is green. It is readily oxidised to form a violet-coloured species. Apparently the reduction product is also violet. Thin-film zinc *bis*(naphthalocyanine) is also green when neutral. Electrochemical oxidation of $Zn(naph)_2$ yields a brown-coloured product, but reduction is not possible [56]. The electrochemical processes accompanying colour change show little reversibility for either electrochrome [56, 57]. A 'triple-decker' naphthalocyanine compound [naph–Lu–naph–Lu–naph] has recently been reported [59].

The electrochromism of the pyridinoporphyrazine system has also received some attention [60]. Here, the ligand is similar to phthalocyanine but with quaternised pyridine occupying all four phenyl sites *cf.* Fig. 5.1. Cobalt at the centre of a single ring has also been studied [60].

References

[1] I.S. Kirin, P.N. Moskalev and Y.A. Makashen, *Russ. J. Inorg. Chem.,* 10 (1965) 1951.

[2] G.S.E. Collins and D.J. Schiffrin, *J. Electroanal. Chem.*, 139 (1982) 335.

[3] G.S.E. Collins and D.J. Schiffrin, *J. Electrochem. Soc.*, 132 (1985) 1835.

[4] M.M. Nicholson and R.V. Galliardi, *S.I.D. Int. Symp. Dig.*, IX (1978) 24.

[5] M.M. Nicholson and F.A. Pizzarello, *J. Electrochem. Soc.*, 126 (1979) 1490.

[6] M.M. Nicholson and F.A. Pizzarello, *J. Electron. Mater.*, 26 (1979) 1490.

[7] M.M. Nicholson and F.A. Pizzarello, *J. Electrochem. Soc.*, 127 (1980) 821.

[8] M.M. Nicholson and F.A. Pizzarello, *J. Electrochem. Soc.*, 127 (1980) 2617.

[9] M.M. Nicholson and F.A. Pizzarello, *J. Electrochem. Soc.*, 128 (1981) 231.

[10] M.M. Nicholson and F.A. Pizzarello, *J. Electrochem. Soc.*, 128 (1981) 1740.

[11] M.M. Nicholson and F.A. Pizzarello, *J. Electrochem. Soc.*, 128 (1981) 1288.

[12] M.M. Nicholson, *Ind. Eng. Chem., Prod. Res. Develop.*, 21 (1982) 261.

[13] Y. Bessonant, G. Gerard and G. Leroy, *Proceedings of the First European Display Research Conference* (München), 1981, 104.

[14] P.N. Moskalev and N.I. Kirina, *J. Appl. Chem. U.S.S.R.*, 48 (1975), 370.

[15] A.T. Chang and J.C. Marchon, *Inorg. Chim. Acta,* 53 (1981) L241.

[16] P.N. Moskalev, G.N. Shapkin and A.N. Darovskikh, *Russ. J. Inorg. Chem.*, 24 (1979) 188.

[17] P.N. Moskalev and G.N. Shapkin, *Sov. Electrochem.*, 14 (1978) 486.

[18] G.A. Corker, B. Grant and N.J. Clecak, *J. Electrochem. Soc.*, 126 (1979) 1339.

[19] V. Plichon, R. Even and G. Beiner, *J. Electroanal. Chem.*, 305 (1991) 195.

[20] P.N. Moskalev and I.S. Kirin, *Russ. J. Inorg. Chem.*, 15 (1970) 7.

[21] A.F. Sammells, *Government Reports and Announcements Index US*, 87 (1987) Abstr. No. 703,869, cited in *Chem. Abstr*. 107: 86,014*m*.

[22] A.F. Sammells and N.U. Pujare, *J. Electrochem. Soc.*, 133 (1986) 1065.

[23] M.M. Nicholson and F.A. Pizzarello, *Government and Reports Announcements Index US,* 88 (1988) absract no. 823,879, cited in *Chem. Abstr.* 110: 48,312*v*.

[24] F. Castaneda and V. Plichon, *J. Electroanal. Chem.*, 236 (1987) 163.

[25] L.R. Faulkner and J.M. Green, *J. Am. Chem. Soc.*, 105 (1983) 2950.

[26] Y. Kohno, M. Masui, K. Ono, T. Wada and M. Takeuchi, *Jpn. J. Appl. Phys.*, 31 (1992) L252.

[27] J. Silver, P. Lukes, P. Hey and M.T. Ahmet, *J. Mater. Chem.*, 2 (1992) 841.

[28] W. Zhang, *Shaghai Jiaotong Daxue Xuebao*, 24 (1990) 67, cited in *Chem. Abstr.* 114 (10) 90,560*g*.

[29] M. Starke, I. Androsche and C. Hamann, *Phys. Status Solidi A*, 120 (1990) K95.

[30] J. Silver, J. Billingham and D.J. Barber, in C. Shi, H. Li and A. Scott (eds.), 'The First Pacific Rim International Conference on Advanced Materials and Processing', The Minerals, Metals and Materials Society, 1992.

[31] J. Silver, P. Lukes, A. Houlton, S. Howe, P. Hey and M.T. Ahmet, *J. Mater. Chem.*, 2 (1992) 849.

[32] L.R. Faulkner, J.L. Kahl, K. Dwarakanath and H. Tackikawa, *J. Am. Chem. Soc.,* 108 (1986) 5438.

[33] J. Silver, P. Lukes, P. Hey and M.T. Ahmet, *J. Mater. Chem.*, 1 (1991) 881.

[34] P. Corbeau, M.T. Riou, C. Clarisse, M. Bardin and V. Plichon, *J. Electroanal. Chem.,* 274 (1989) 107.

[35] M. Petty, D.R. Lovett, J. Miller and J. Silver, *J. Mater. Chem.*, 1 (1991) 971.

[36] B. Lukas, D.R. Lovett and J. Silver, *Thin Solid Films*, 210/211 (1992) 213.

[37] J. Muto and K. Kusayanagi, *Phys. Status Solidi A,* 126 (1991) K129.

[38] J. Mark and L.R. Faulkner, *J. Am. Chem. Soc.*, 105 (1983) 2950.

[39] J. Silver, P. Lukes, S.D. Howe and B. Howlin, *J. Mater. Chem.*, 1 (1991) 29.

[40] C.S Frampton, J.M. O'Connor, J. Peterson and J. Silver, *Displays*, 9 (1988) 174.

[41] D. Walton, B. Ely and G. Elliott, *J. Electrochem. Soc.*, 128 (1981) 2479.

[42] J.T.S. Irvine, B.R. Eggins and J. Grimshaw, *J. Electroanal. Chem.*, 271 (1989) 161, and references therein.

[43] C.C. Leznoff, H. Lam, S.M. Marcuccio, W.A. Nevin, P. Janda, N. Kobayashi and A.B.P. Lever, *J.C.S., Chem. Commun.*, (1987) 699.

[44] W.A. Nevin, M.R. Hempstead, W. Liu, C.C. Leznoff and A.B.P. Lever, *Inorg. Chem.*, 26 (1987) 570.

[45] W.A. Nevin, W. Liu, S. Greenberg, M.R. Hampstead, S.M. Marcuccio, M. Melnik, C.C. Leznoff and A.B.P. Lever, *Inorg. Chem.*, 26 (1987) 891.

[46] W.A. Nevin, W. Liu and A.B.P. Lever, *Can. J. Chem.*, 65 (1987) 855.

[47] W.A. Nevin, W. Liu, M. Melnik and A.B.P. Lever, *J. Electroanal. Chem.*, 213 (1986), 217.

[48] M. Yamana, K. Yanda, N. Kashiwazaki, M. Yamamoto and C. Walton, *Jpn. J. Appl. Phys.*, 28 (1989) L1592.

[49] N. Kashiwazaki, *Solar Energy Mater. Solar Cells*, 25 (1992) 349.

[50] N. Kashiwazaki and M. Yanana, 12th International Display Research Conference, Hiroshima, Japan, 1992. Abstract number P3–19

[51] N. Kashiwazaki, *Jpn. J. Appl. Phys.*, 31 (1992) 1892.

[52] N. Kashiwazaki, *Electronics and Communications in Japan*, 75 (1992) 67.

[53] H. Yanagi, K. Takeshita and M. Ashida, *Defect Control in Semiconductors*, 2 (1990) 1635, cited in *Chem. Abstr.* 114 (21) 206,456*j*.

[54] K. Ohta, T. Fujimoto, T. Komatsu, N. Maeda, N. Koike, N. Fujimori and I. Yamamoto, *Kidorui*, 16 (1990) 124, cited in *Chem. Abstr.* 114 (20) 196, 866 *b*.

[55] D. Schlettwein, M. Kaneko, A. Yamada, D. Wöhrle and N.I. Jaeger. *J. Phys. Chem.*, 95 (1991) 1748.

[56] H. Yanagi and M. Toriida, *Proc. Electrochem. Soc.*, 94–2 (1994) 211.

[57] H. Yanagi and M. Toriida, *J. Electrochem. Soc.*, 141 (1994) 64.

[58] M. Shimura, K. Satoh, I. Idenuma and Y. Shimura, *Hyoumengijutsu*, 41 (1990) 938 cited in reference [56] here.

[59] F. Guyon, A. Pondaven and M. L'Her, *J.C.S., Chem. Commun.*, (1994) 1125.

[60] Y. Yamada, N. Kashiwazaki, M. Yamamoto and T. Nakano, *Displays*, 9 (1988) 190.

6 Prussian Blue: Its Systems and Analogues

6.1 Introduction: Historical and Bulk Properties

Prussian blue (PB, ferric ferrocyanide or iron(III) hexacyanoferrate(II)) – the original photochromic, its image application a part of common parlance* – first made by Diesbach in Berlin in 1704 [1] has had a long history both in coordination chemistry and as a pigment used extensively in the formulation of paints, lacquers, and printing inks [2–5]. It has recently featured as a much-loved garment colour in novels of a best-selling authoress [6]. The first report in 1978 of the electrochemistry and electrochromism of PB films is relatively recent [7]. Following this, numerous studies concerning the electrochemistry of PB and related analogues have been made and the subject has been reviewed [8].

PB is the fore-runner of a number of polynuclear transition metal hexacyanometallates which form an important class of insoluble mixed valence compounds [9–11]. They have the general formula $M'_k[M''(CN)_6]_l$ (k, l integral) where M' and M'' are transition metals with different formal oxidation numbers. These materials can contain ions of other metals and varying amounts of water. In PB the two transition metals in the formula are the two common oxidation states of iron, Fe^{III} and Fe^{II}. Preparation of PB can easily be demonstrated by mixing aqueous solutions of either a ferricyanide salt with a ferrous salt or a ferrocyanide salt with a ferric salt (the latter product formally known as Turnbull's blue). In the PB chromophore, the distribution of oxidation states is Fe^{III}–Fe^{II} respectively, in $Fe^{3+}Fe^{II}(CN)_6^{4-}$, which was established by the C≡N stretch frequency in the IR spectrum and confirmed by Mössbauer spectroscopy [12]; Fe^{3+} is high spin and can have H_2O attached; Fe^{II} is low-spin.

While the composition in detail of the PB solids is extraordinarily preparation-sensitive, the major classification of extreme cases delineates 'insoluble' PB (*i*-PB) $Fe^{3+}[Fe^{3+}Fe^{II}(CN)_6^{4-}]_3$ and 'soluble' PB (*s*-PB), $K^+Fe^{3+}Fe^{II}(CN)_6^{4-}$. The epithets are misnomers, referring to ease of peptisation rather than dissolution, since both are intrinsically very insoluble ($K_{sp} \sim 10^{-40}$). The $Fe^{3+}Fe^{II}(CN)_6^{4-}$ chromophore falls into Group II of the Robin-Day mixed-valence classification (section 4.1), the blue intervalence CT band on analysis of the intensity indicating ~1% delocalisation of the transferable electron in the ground state (i.e. before any optical CT) [13].

---

* Chambers's Twentieth Century Dictionary, Ed. W. Geddie, 1954, defines 'blueprint' as "a photographic print, white upon blue, on paper sensitised with a ferric salt and potassium ferricyanide from a photographic negative or a drawing on transparent paper – also called cyanotype, ferroprussiate print: a preliminary sketch or plan of work to be done". (Colouration of pale-yellow ferric-ferricyanide ensues from photoreduction to Prussian blue, involving included H_2O.)

X-ray powder diffraction patterns for *s*-PB indicate a face-centred cubic lattice, with the high-spin Fe^{III} and low-spin Fe^{II} ions coordinated octahedrally by –N≡C and –C≡N ligands respectively, K^+ ions occupying interstitial sites [14]. In *i*-PB, Mössbauer spectroscopy confirms the interstitial ions to be Fe^{3+} [12]. The *single crystal* X-ray diffraction patterns of Ludi *et al.* [15] indicate however a primitive cubic lattice, where 1/4 of the Fe^{II} sites are vacant. This proposed structure contains no interstitial ions, with 1/4 of the Fe^{III} centres being coordinated by 6 –N≡C ligands, the remainder by 4 –N≡C, and every Fe^{II} centre surrounded by 6 –C≡N ligands. The absence of certain lines in the single-crystal X-ray data suggests that the Fe^{II} vacancies are randomly distributed, and occupied by water molecules which complete the octahedral coordination about Fe^{III}. The widespread assumption of Ludi's model for *i*-PB may be questioned in view of the probably substantial differences between Ludi's slowly grown single crystals and usually polycrystalline rapid growths.

6.2 Preparation of Prussian Blue Thin Films

PB thin films are generally prepared by the original method based on electrochemical deposition [7], although electroless deposition [16], and sacrificial anode (SA) methods [17, 18] have been described. Thus PB films can be electrochemically deposited onto a variety of electrode substrates by electroreduction of solutions containing iron(III) and hexacyanoferrate(III) ions. PB electrodeposition has been studied by numerous techniques. Voltammetry [19–21] and galvanostatic studies [22] have shown that reduction of iron(III)hexacyanoferrate(III) is the principal electron transfer process in PB electrodeposition. This brown/yellow soluble complex is present in solutions containing iron(III) and hexacyanoferrate(III) ions as a result of the equilibrium:

$$Fe^{3+} + [Fe^{III}(CN)_6]^{3-} \rightleftharpoons [Fe^{III}Fe^{III}(CN)_6]^0 \qquad (6.1)$$

Chronoabsorptiometric studies [23] for galvanostatic PB electrodeposition onto ITO electrodes have shown that the absorbance due to the CT band of the growing PB film is proportional to the charge passed. Quartz-crystal microbalance measurements [24] for potentiostatic PB electrodeposition onto gold have revealed that the mass gain per unit area is proportional to the charge passed. The calculated molar mass indicates that the PB films are highly hydrated. Ellipsometric measurements [25] for potentiostatic PB electrodeposition onto platinum showed that the level of hydration was around 34 H_2O per PB unit cell. Changes in the ellipsometric parameters during PB electrodeposition revealed growth of a single homogeneous film for the first 80 s followed by growth of a second, outer, more porous film on top of the relatively compact inner film. Chronoamperometric measurements (over a scale of several seconds) supported by SEM for the electrodeposition of PB onto ITO and platinum by electroreduction from iron(III) hexacyanoferrate(III)-containing solutions have been performed [26]. Variation of electrode potential, supporting electrolyte and electroactive species concentration have established a three-stage electrodeposition mechanism. The main features of the mechanism which are inferred from the chronoamperometry are as follows:

- In the first growth phase the surface becomes uniformly covered as small PB nuclei form and grow on electrode substrate sites. This results in a decrease in the number of sites at which direct electroreduction can occur and a decrease in the film formation rate.
- In the second growth phase there is an increase in rate towards maximal roughness as the electroactive area increases by formation and three-dimensional growth of PB nuclei attached to the PB interface formed in the initial stage. These processes are evident in the SEM.
- In the final growth phase, diffusion of locally depleted electroactive species to the now three-dimensional PB interface plays an increasingly dominant role and limits electron transfer resulting in a fall in rate.

The electrochemical deposition method described above has been used extensively, especially on the laboratory scale. However, Ho [18] has noted that this method is not suitable for forming PB films over a large surface area because of the non-uniform film thickness which accompanies the potential drop across the conducting substrate. Although a uniform electroless deposition method using the reducing agent H_3PO_2 has been proposed to overcome this drawback [16], it takes more than five hours to grow the same amount of PB as is formed by the electrodeposition method. On the other hand, the sacrificial anode method, proposed by Ellis *et al.* [17], is a relatively uniform and fast deposition method without these drawbacks. Ho believes that the sacrificial anode method is the most promising one for easy scale-up and better control of thickness uniformity. However, it is not clear why spontaneous operation with a sacrificial anode as counter electrode is better than an inert counter electrode and driven electron transfer; Ho's improvement may thus have arisen simply from better electrode dispositions, or from the control of deposition rate by the anode dissolution rate.

6.3 Prussian Blue Electrochromic Films: Cyclic Voltammetry, *In Situ* Spectroscopy and Characterisation

Electrodeposited PB films may be partially oxidised [19–21] to Prussian green (PG), a species historically known as Berlin green (BG):

$$\underset{\text{(PB)}}{[Fe^{III}Fe^{II}(CN)_6]^-} \rightarrow \underset{\text{(PG)}}{[Fe^{III}\{Fe^{III}(CN)_6\}_{2/3}\{Fe^{II}(CN)_6\}_{1/3}]^{1/3-}} + {}^{2}/_{3}\, e^- \quad (6.2)$$

where the fractions $^1/_3$ and $^2/_3$ are illustrative rather than precise. Thus, although in bulk form PG is believed to have a fixed composition with anion composition as above, it has been demonstrated that there is a continuous composition range between PB and PX for thin films [21]. Fully oxidised PB is Prussian brown (PX), which appears brown as a bulk solid, brown/yellow in solution, and golden yellow as a particularly pure form that is prepared on electro-oxidation of thin-film PB [20, 21]:

$$\underset{\text{(PB)}}{[\mathrm{Fe^{III}Fe^{II}(CN)_6}]^-} \rightarrow \underset{\text{(PX)}}{[\mathrm{Fe^{III}Fe^{III}(CN)_6}]^0} + \mathrm{e^-} \qquad (6.3)$$

Reduction of PB yields Prussian white (PW), also known as Everitt's salt, which appears colourless as a thin film:

$$\underset{\text{(PB)}}{[\mathrm{Fe^{III}Fe^{II}(CN)_6}]^-} + \mathrm{e^-} \rightarrow \underset{\text{(PW)}}{[\mathrm{Fe^{II}Fe^{II}(CN)_6}]^{2-}} \qquad (6.4)$$

For all redox reactions above there is concomitant ion movement into/out of the films for electroneutrality.

The CV for a typical PB-modified electrode is shown in Fig. 6.1, together with a picture of one of the authors (DRR) and his charming assistant viewing their demonstration cell undergoing the various colour changes. Whilst *s*-PB, *i*-PB, PG and PW are all insoluble in water, PX is slightly soluble in its pure (golden yellow) form (indeed the electrodeposition technique depends on the solubility of the $[\mathrm{Fe^{III}Fe^{III}(CN)_6}]^0$ complex). This implies a positive potential limit of about +0.9 V for a high write–erase efficiency in contact with water. Although practical PB ECDs have primarily exploited the PB $\rightleftharpoons$ PW transition, this does not rule out the prospect of four-colour PB poly-electrochromic ECDs, as other solvent systems may not dissolve PX.

The spectra of the yellow, green, blue, and clear ('white') forms of PB and its redox variants are shown in Fig. 6.2, together with two intermediate states between the blue and the green. The yellow absorption band corresponds with that of $[\mathrm{Fe^{III}Fe^{III}(CN)_6}]^0$ in solution, both maxima being at 425 nm and coinciding with the (weaker) $[\mathrm{Fe^{III}(CN)_6}]^{3-}$ absorption maximum. On increase from +0.50 V to more oxidising potentials, the original 690 nm PB peak continuously shifts to longer wavelengths with diminishing absorption, while the peak at 425 nm steadily increases, owing to the increasing $[\mathrm{Fe^{III}Fe^{III}(CN)_6}]^0$ absorption. The reduction of PB to PW is by contrast abrupt, with transformation to all PW or all PB without pause, depending on the potential set. The broad CV peak for PB $\rightleftharpoons$ PX in contrast with the sharp PB $\rightleftharpoons$ PW transition emphasises the range of compositions involved. This difference in behaviour, supported by ellipsometric measurements [25], indicates continuous mixed-valence compositions over the blue-to-yellow range in contrast with the presumably immiscible PB and PW which clearly transform one into the other without intermediacy of composition.

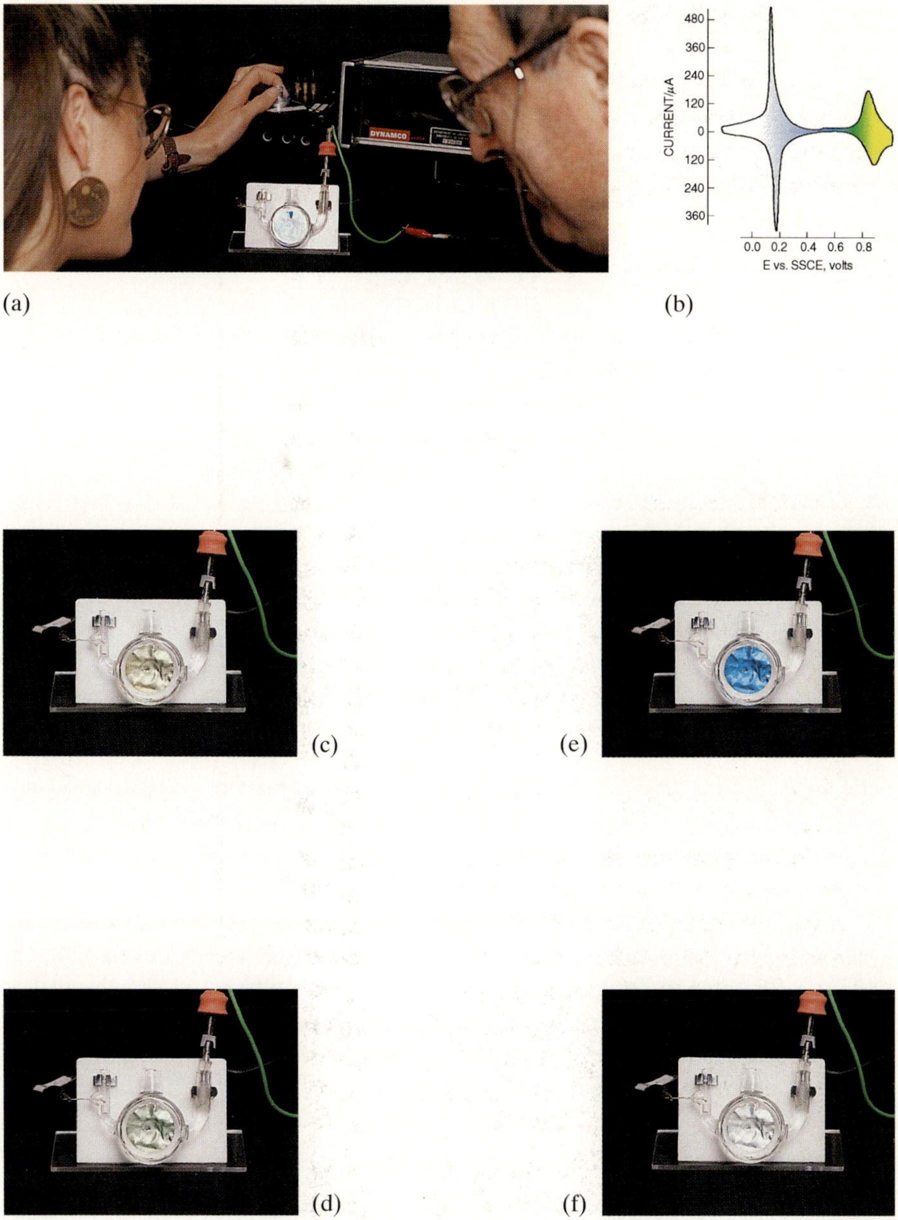

Fig. 6.1 *Prussian blue: colours with potential (a) Author (D.R.R.) and Sandra Mann examine cell, (b) The CV, starting at 0.5 V then proceeding cathodically, (c) +1.0 V, (d) +0.8 V, (e) +0.5 V, (f) -0.2 V.*

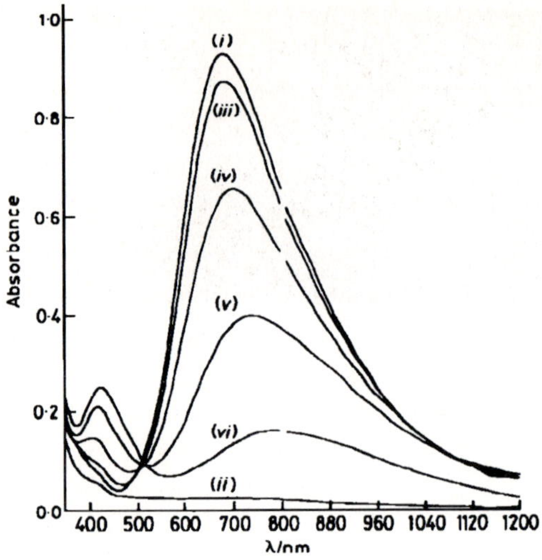

Fig. 6.2 Spectra of PB films at various potentials [(i) +0.50, (ii) –0.20, (iii) +0.80, (iv) +0.85, (v) +0.90, and (vi) +1.20 V (vs. SCE)] with 0.2 M KCl + 0.01 M HCl as supporting electrolyte. After electrodeposition, the PB/ITO electrode was cycled in 2 M KCl, +0.50 to –0.20 V at 10 mV s^{-1}, prior to the spectroelectrochemical measurements described. (Figure reproduced from ref. [21] with permission of the Royal Society of Chemistry.)

The identity as *s*-PB or *i*-PB of the initially electrodeposited PB has been debated in the literature [21, 27–32]. Based on changes that take place in the intervalence CT band on redox cycling it has been postulated that *i*-PB is first formed, followed by a transformation to *s*-PB on potential cycling [21]. Further evidence is the difference in the CV response for the PB $\rightleftharpoons$ PW transition between the first cycle and all succeeding cycles, suggesting structural reorganisation of the film during the first cycle [20]. On soaking *s*-PB films in saturated $FeCl_3$ solutions partial reversion of the absorbance maximum and broadening of the spectrum, approaching the values observed for *i*-PB, is found [21]. Itaya and Uchida [27] however claim that the film is always *i*-PB, their argument being based on the ratio of charge passed on oxidation to PX to that passed on reduction to PW, which was 0.708 rather than 1.00, that is, equations (6.5) and (6.6) are applicable:

$$\underset{(i\text{-PB})}{Fe^{3+}[Fe^{III}Fe^{II}(CN)_6]_3} + 4e^- + 4K^+ \rightarrow \underset{(\text{PW})}{K_4Fe^{2+}[Fe^{II}Fe^{II}(CN)_6]_3} \qquad (6.5)$$

$$\underset{(i\text{-PB})}{Fe^{3+}[Fe^{III}Fe^{II}(CN)_6]_3} - 3e^- + 3X^- \rightarrow \underset{(\text{PX})}{Fe^{3+}[Fe^{III}Fe^{III}(CN)_6]_3X_3} \qquad (6.6)$$

Emrich *et al.* [28] using XPS data, and Lundgren and Murray [29] with CV, EDAX, XPS, elemental analysis and spectroelectrochemical observations have confirmed *i*-PB as the first deposited form, with a *gradual* transformation to *s*-PB on potential cycling. Other support for the *i*-PB to *s*-PB transformation comes from the ellipsometric study by Ord *et al.* [30] who found that the PB film *after* the first and subsequent cycles for the PB $\rightleftarrows$ PW transition had optical properties which differ from the original PB film. Results from *in situ* Fourier-transform infra red (FTIR) spectroscopy give further support to an *i*-PB to *s*-PB transformation on repeated reductive cycling [32]. Quartz-crystal microbalance mass change measurements on voltammetrically scanned PB films reinforce the theory of lattice reorganisation during the initial film reduction [24].

Whilst PB film stability is frequently discussed in the preceding papers, Stilwell *et al.* [33] have studied in detail the factors that influence the cycle stability of PB films. They found that electrolyte pH was the overwhelming factor in film stability; cycle numbers in excess of 100,000 were easily achieved in solutions of pH 2–3. Concurrent with the increase in stability at lower pH was a considerable increase in switching rate. Films grown from chloride-containing solutions were found to be slightly more stable, in terms of cycle life, compared to those grown from chloride-free solutions. In addition, they found that there occurs at least some conversion of PB form during the first cycle.

6.4 Prussian Blue ECDs

6.4.1 ECDs with Prussian Blue as Sole Electrochrome

Itaya *et al.* [34] and DeBerry and Viehbeck [35] were first to demonstrate the use of PB in display devices. Itaya *et al.* described a seven-segment ECD display using PB-modified SnO_2 working and counter electrodes at 1 mm separation, with TiO_2 powder as a diffuse scattering background. The supporting electrolyte of the ECD, although not detailed was likely to be 1 M KCl (pH 4.0 HCl) as used for traditional PB-modified electrode CV and pulsing experiments also described in the report. DeBerry and Viehbeck [35] described a light addressable photoelectrochromic electrode based on the photo-initiated oxidation of PW. This device consisted of a PB-modified anodised Ti sheet, ITO counter-electrode and Ag/AgCl reference electrode. In operation, PB is first reduced to PW by polarisation to −0.60 V. In the dark, with *n*-TiO_2 as substrate, this process cannot be reversed. However, imaging can take place on UV light exposure at an electrode potential of 0.15 V. Further studies on this system have been made [36, 37]. The phenomenon of photoelectrochromism is considered more generally in chapter 12.

Honda *et al.* [38–40], whose primary motivation was the demonstration of a novel rechargeable battery, first demonstrated the use of solid polymer electrolyte in the construction of PB ECDs. Device fabrication involved immersion of a Nafion® (sulphonated polytetrafluoroethane polymer) 117 membrane (160 μm × 1 cm × 1 cm) in

50 mM $FeCl_2$ aqueous solution for 1 h under argon. The membrane, after washing and drying, was then immersed for 15–20 h in 50 mM $K_3Fe(CN)_6$ aqueous solution. The resulting PB-containing Nafion® composite film, after removal of unreacted metal ions, was sandwiched between two ITO plates to prepare a solid-state cell. Electron-probe microanalysis and optical microscopy revealed a phase-separated structure with PB formation occurring within 20–40 μm layers at each surface of the membrane. Honda *et al.* noted that the redox behaviour of the ITO | PB-Nafion® | ITO solid-state cell would be expected to occur at a potential that is equal to the difference in potential of the redox reactions occurring at a single film PB-modified electrode. Thus green-to-blue electrochromicity occurred at ±0.68 V. In a report highlighting the electrochromic properties of the above device [40], a white reflection backboard was incorporated into the solid polymer electrolyte by coating Nafion® on a microporous polyethylene film ('Hipore' from Asahi Chemical Industry, 50 μm thickness and pore size 0.5 μm). The response time of this device was around 2 s. The inclusion of the white microporous film as reflection backboard increased the contrast ratio 4–5 times over that of the device without backboard.

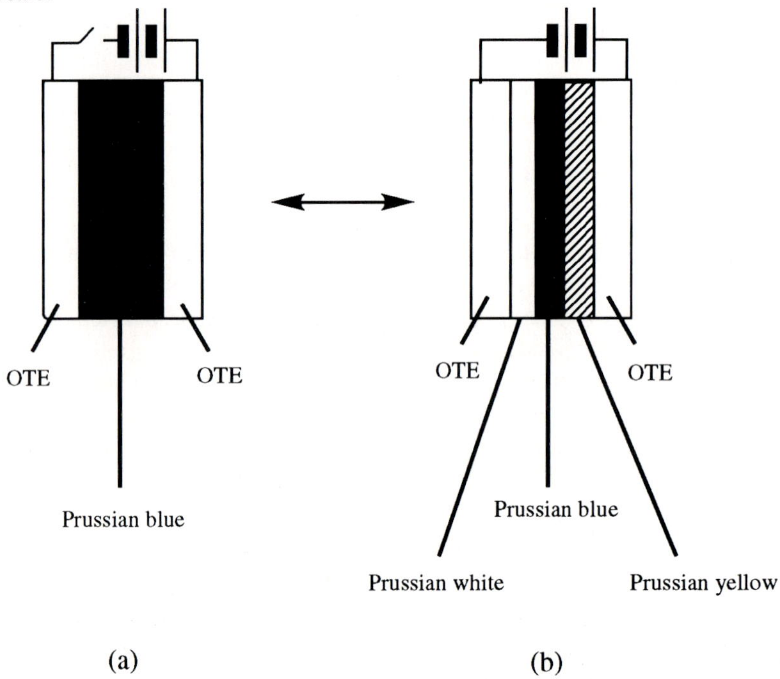

Fig. 6.3 Schematic cross section of a single film PB cell. The cell is shown (a) without and (b) with a voltage applied. After ref. [41].

The construction and optical behaviour of an ECD utilising a *single film* of PB has been described by Carpenter and Conell of General Motors [41]. Light transmission through the device can be increased by application of 2–12 V across the film; the larger voltages bleach more of the PB, to a maximum of about 50%, forming PX (yellow) at the positive and PW (clear) at the negative electrode. The device is fully coloured in the 'off' state and has no optical memory. The device described is novel in that no conventional electrolyte is added. In the design, a film of PB is sandwiched between transparent conducting plates as shown in Fig. 6.3(a). Upon application of an appropriate potential across the film, oxidation occurs near the positive electrode and reduction near the negative electrode to yield PX and PW respectively, as shown schematically in Fig. 6.3(b). The conversion of the outer portions of the film results in a net bleaching of the device. The functioning of the device relies on two properties of PB: (i) it can be bleached both anodically and cathodically, and (ii) it is a mixed conductor through which potassium ions can move to provide charge compensation required for the electrochromic redox reactions. Thus, in the device, a single film of PB is made to function as both primary and complementary electrodes and also as the electrolyte. Drawbacks however include the lack of optical memory in contrast to other ECDs, the applied voltage determining the extent of bleaching. Furthermore, the response is sensitive to the pressure applied to the film and to the extent of film hydration.

6.4.2 Prussian-Blue – Tungsten-Trioxide ECDs

Numerous workers [42–56] have combined WO_3 and PB in ECDs that exhibit deep blue to transparent electrochromicity. Since PB is an anodically-colouring electrochrome, and WO_3 a cathodically-colouring electrochrome (see chap. 4), they can be used together in a single device so that their electrochromic reactions are complementary:

$$\underset{\text{(blue)}}{[Fe^{III}Fe^{II}(CN)_6]^-} + e^- \rightarrow \underset{\text{(colourless)}}{[Fe^{II}Fe^{II}(CN)_6]^{2-}} \tag{6.7}$$

$$\underset{\text{(very pale yellow)}}{WO_3 + x(M^+ + e^-)} \quad \rightarrow \quad \underset{\text{(blue)}}{M_x \overset{VI}{W}_{(1-x)} \overset{V}{W}_x O_3} \tag{6.8}$$

The construction of such a device is shown in Fig. 6.4; thin films of these materials deposited on OTE electrodes are separated by a layer of a transparent ionic conductor. The films can be coloured simultaneously when a sufficient voltage is applied between them such that the WO_3 electrode is the cathode and the PB electrode the anode. Conversely, the coloured films can be bleached to transparency when the polarity is reversed, returning the ECD to a transparent state. Kase, Kawai and Ura of the Nissan Motor Co. Ltd. were first to demonstrate the combination of WO_3 and PB in a complementary ECD, with the aim of producing adjustable transparency glass for automotive vehicle applications [42].

A 400 mm × 400 mm device was demonstrated with 1 M $LiClO_4$ | propylene carbonate as supporting electrolyte. More recent devices have used solid polymer electrolytes. Ho, Rukavina and Greenberg [46] have tabulated a partial list of complementary ECDs, the WO_3 | PB systems from which are shown in Table 6.1. Their own devices (with poly-AMPS, the proton-conducting polymer electrolyte) operate at low applied voltages, +1.2 V to darken and –0.6 V to bleach [46]. They claim colour switching over 20,000 cycles without significant degradation or irreversible side reactions. The sustained high overall colouration efficiency of the devices is believed to confirm the insertion/extraction of protons into and out of both WO_3 and PB films.

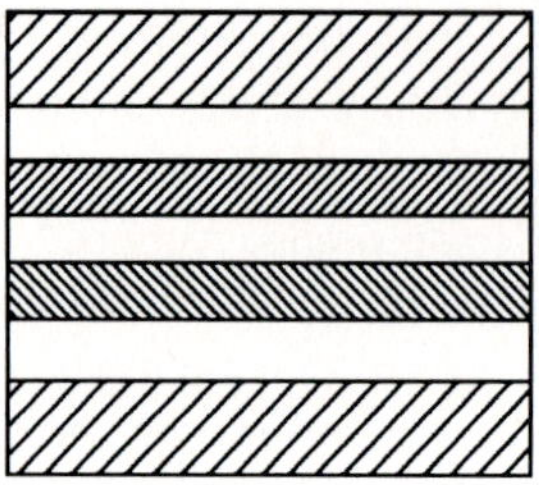

Fig. 6.4 Schematic diagram of a complementary WO_3–PB ECD with a solid polymer electrolyte. After ref. [46].

Table 6.1 A Partial List of Tungsten-oxide–PB Complementary ECDs. From ref. [46].

Configuration	η/cm^2 C^{-1}	λ_{max}/nm	Ref.
K_xWO_3 | H_3PO_4–KH_2PO_4 in PVA | PB	127.7	690	[45]
H_xWO_3 | PolyAMPS | PB	75.1	550	[46]
K_xWO_3 | KCF_3SO_3 in PEO–PU | PB	145	632.8	[47–49]
Li_xWO_3 | Li^+-OMPE | PB	105.8	790	[50]
Li_xWO_3 | $LiCF_3SO_3$ in PAA–PEO | PB	—	—	[51]
Li_xWO_3 | $LiClO_4$ in PC | PB	59.0	vis	[52–55]

Key: PEO = poly(ethylene oxide); PU = polyurethane; PVA = poly(vinyl alcohol); OMPE = oxymethylene poly(oxyethylene); PAA = poly(acrylic acid); PC = propylene carbonate; polyAMPS = poly(2-acrylamido-2-methylpropanesulphonic acid).

6.4.3 Prussian-Blue – Polyaniline ECDs

Numerous workers [57–62] have combined PB with the conducting polymer polyaniline (polyaniline) in complementary ECDs that exhibit deep blue-to-light green electrochromism. Electrochromic compatibility is obtained by combining the coloured oxidised state of the polymer (see chap. 9) with the blue PB and the bleached reduced state of the polymer with PG:

$$\underset{\text{(coloured)}}{\text{Oxidised polyaniline + PB}} \rightarrow \underset{\text{(bleached)}}{\text{Emeraldine polyaniline + PG}} \qquad (6.9)$$

Mastragostino *et al.* [57, 58] have described both liquid electrolyte and solid-state configurations. In the former [57], an ECD was assembled by facing off polyaniline | ITO (oxidised at +0.60 *vs.* SCE) and PB | ITO electrodes at 2 mm separation in 1.0 M Na_2SO_4 | H_2SO_4 aqueous supporting electrolyte. The polyaniline (with a cellulose acetate coating for enhanced polyaniline adhesion) and PB films were electrochemically deposited on the ITO electrodes with quantitative charge balance. In a *solid-state* device Mastrogostino *et al.* [58] have used as electrolyte $LiClO_4$-doped Hydrin, a commercially available elastomer, that is prepared by copolymerisation of ethylene oxide with epichlorohydrin. The solid electrolyte was cast on the polyaniline | ITO and PB | ITO electrodes from a tetrahydrofuran solution containing the elastomer and $LiClO_4$. The ECD was assembled by gluing both parts using the polymer electrolyte as adhesive, and making the electrical contacts. The device was characterised by studying its behaviour under repeated double step spectrochronoamperometric experiments using 650 nm monochromatic light. The fall in electrochromic efficiency for the bleaching process from initially 86.7 $cm^2\ C^{-1}$ to 44.3 $cm^2\ C^{-1}$ after 3.5×10^3 pulses was interpreted as being due to the low coulombic efficiency of the cell. Although this aspect was disappointing, the optical memory of the device under open circuit was good for the coloured form, which showed no spectral variation after 10 h, indicating that the oxidised form of polyaniline and PB are stable. However the bleached form showed a pronounced spectral variation in the first 10 min, becoming coloured after 6 h.

Jelle and Hagen [60–62] have recently developed an electrochromic window for solar modulation using PB, polyaniline *and* WO_3. Leventis and Chung [63] had earlier electrochemically deposited PB films onto polyaniline coatings and found that the polyaniline film greatly enhanced the cycle life of PB (in $KHSO_4$ solution), while PB itself enhanced the colouration. Jelle and Hagen took advantage of this *symbiotic* relationship between polyaniline and PB, and incorporated PB together with polyaniline and WO_3 in a complete solid-state electrochromic window. The total device comprised Glass | ITO | polyaniline | PB | polyAMPS | WO_3 | ITO | Glass. Compared with their earlier results with a polyaniline | WO_3 window, Jelle and Hagen were able to block off much more of the light by inclusion of PB within the polyaniline matrix, while still regaining about the same transparency during the bleaching of the window.

6.4.4 A Prussian-Blue – Ytterbium-*bis*(phthalocyanine) ECD

Kashiwazaki [64] has fabricated a complementary ECD using plasma-polymerised ytterbium *bis*(phthalocyanine) (pp–Yb$(pc)_2$) and PB films on ITO with an aqueous solution of 4 M KCl as electrolyte. Blue-to-green electrochromicity was achieved in a two-electrode cell by complementing the green-to-blue colour transition (on reduction) of the pp–Yb$(pc)_2$ film with the blue (PB)-to-colourless (PW) transition (oxidation) of the PB. A three-colour display (blue, green and red) was fabricated in a three-electrode cell in which a third electrode (ITO) was electrically connected to the PB electrode. A reduction reaction at the third electrode, as an additional counter electrode, provides adequate oxidation of the pp–Yb$(pc)_2$ electrode, resulting in the red coloration of the pp–Yb$(pc)_2$ film.

6.5 Prussian Blue Analogues

The PB analogues (other polynuclear transition metal hexacyanometallates [9–11]) that have been reported as thin films are surveyed in this section. Whilst the majority are expected to be electrochromic, absorption spectra as a function of redox state have rarely been reported, and only ruthenium purple and osmium purple have been used in prototype ECDs. The field therefore appears to be open for further investigation and exploitation.

6.5.1 Ruthenium Purple and Osmium Purple

Bulk ruthenium purple (RP, ferric ruthenocyanide or iron(III) hexacyanoruthenate(II)) is synthesised via precipitation from solutions of the appropriate iron and hexacyano-ruthenate salts. The visible absorption spectrum of a colloidal suspension of bulk synthesised RP with potassium as counter cation confirms ferric ruthenocyanide as chromophore [65]. The X-ray powder pattern with iron(III) as counter cation gives a lattice constant of 10.42 Å as compared to 10.19 Å for the PB analogue [66]. However, although no single-crystal studies have been made, RP could have a disordered structure similar to that reported for a PB single crystal [15]. The potassium and ammonium salts give cubic powder patterns similar to their PB analogues [67].

RP films have been prepared by electroreduction of the soluble ferric ruthenicyanide complex either potentiostatically [68], galvanostatically [69, 70] or using a copper wire as sacrificial anode [68]. The visible absorption spectrum of RP prepared in the presence of excess potassium ion showed a broad CT band, as for bulk synthesised RP, with a maximum at approximately 550 nm [68]. RP films can be reversibly reduced to the colourless iron(II) hexacyanoruthenate(II) form, although electrooxidation to the Prussian green analogue is not observed. A large background oxidation current is observed in chloride-containing electrolyte, suggesting electrocatalytic activity of RP for either oxygen or chlorine evolution [69].

Ataki *et al.* [70], have prepared PB, RP and osmium purple (OP, ferric osmocyanide or iron(III) hexacyano-osmate(II)) films using the galvanostatic technique of Itaya *et al.* [69]. A seven-segment electrochromic display was described using either PB, RP or OP as the electrochrome. The current efficiency for deposition of RP and OP was considerably less than the 100% obtained in the case of PB. For OP, lower current densities (5 compared to 10 $\mu A\ cm^{-2}$) were necessary in order to obtain uniform films with high current efficiency.

6.5.2 Vanadium Hexacyanoferrate

Vanadium hexacyanoferrate (VHCF) films have been prepared on Pt or FTO by potential cycling from a solution containing Na_3VO_4 and $K_3Fe(CN)_6$ in 3.6 M H_2SO_4 [71–73]. Carpenter *et al.* [71], by correlation with CVs for solutions containing only one of the individual electroactive ions, have proposed that electrodeposition involves the reduction of the dioxovanadium ion VO_2^+ (the stable vanadium(V) ion under these acidic conditions), followed by precipitation with ferricyanide ion. While the reduction of the ferricyanide ion in solution probably also occurs when the electrode is swept to more negative potentials, this reduction does not appear to be critical to film formation, since VHCF films can be successfully deposited by potential cycling over a range positive of that required for ferricyanide reduction. No evidence was obtained for the formation of a vanadium-ferricyanide type complex analogous to ferric ferricyanide, the visible absorption spectrum of the deposition solution being a simple summation of spectra from the single-component solutions. While VHCF films are visually electrochromic, switching from green in the oxidised state to yellow in the reduced state, Carpenter *et al.* [71] show that most of the electrochromic modulation occurs in the UV region. From electrochemical data and XPS they conclude that the electrochromicity involves only the iron centres in the film. The vanadium ions, found to be present predominantly in the +4 oxidation state, are unaffected by the potential.

6.5.3 Nickel Hexacyanoferrate

The synthesis of PB analogue films does not necessarily require electroreduction of soluble solution complexes. For example, Bocarsly and co-workers [74–84] have prepared nickel hexacyanoferrate films by oxidation of nickel in the presence of ferricyanide ions. Indeed a variety of $[Fe(CN)_5L]^{n-}$ complexes can be attached to nickel electrodes by oxidation in the presence of the appropriate complex. Unlike PB, the corresponding bulk materials do not possess low energy intervalent CT bands and thus are not highly coloured. However, electrochromic surfaces can be generated by substitution of two iron bound cyanides by a suitable bidentate ligand [79]. Thus 2,2'-bipyridine can be indirectly attached to the nickel electrode *via* a cyano-iron complex. When bipyridine is employed as the chelating agent, the parent complex, $[Fe^{II}(CN)_4bpy]^{2-}$, takes on an intense red colour associated with a metal-to-ligand CT (MLCT) visible absorption band centred at 480 nm. This optical transition is sensitive to both the iron oxidation state (only being present in the Fe(II) complex) and the environment of the cyanide nitrogen lone pair. Reaction of the complex with Ni^{2+} either under bulk conditions or at a nickel electrode surface generates a bright red material. By analogy with the parent iron complex this red colour is associated with the $(d\pi)Fe(II) \rightarrow (\pi^*)bpy$ CT transition. For bulk samples, chemical oxidation to the Fe(III) state yields a light-orange material, while modified electrodes can be reversibly cycled between the intensely red and clear forms, a process which correlates well with the observed CV response.

6.5.4 Copper Hexacyanoferrate

Siperko and Kuwana [85–88] have studied copper hexacyanoferrate (CuHCF) films prepared voltammetrically by electroplating a thin film of copper on GC or ITO electrodes in the presence of ferrocyanide ions. Films were deposited by first cycling between +0.40 and +0.05 V in a solution of cupric nitrate in aqueous $KClO_4$. Copper was then deposited on the electrode by stepping the potential from +0.03 to –0.50 V, and subsequently removed (stripped) by linearly scanning the potential from –0.50 to +0.50 V. The deposition and removal sequence was repeated until a reproducible CV was obtained during the stripping procedure. The CuHCF film was then formed by stepping the electrode potential in the presence of cupric ion from +0.03 to –0.50 V followed by injection of an aliquot of $K_4Fe(CN)_6$ solution (a red/brown ferrocyanide sol is formed immediately) into the cell. Other workers have prepared CuHCF films by modification of this original method [89–91]. The CuHCF film formation mechanism has not been elucidated but the co-deposition of copper is important in the formation of stable films. Films formed by galvanostatic or potentiostatic methods from solutions of cupric ion and ferricyanide ion showed noticeable deterioration within a few CV scans. The co-deposition procedure provides a fresh copper surface for film adhesion and the resulting films are able to withstand ~1000 voltammetric cycles. Such scanning of a CuHCF film in 0.5 M K_2SO_4 gave a well-defined reversible couple at *ca.* +0.69 V, characteristic of an adsorbed species.

CuHCF films exhibit red/brown-to-yellow electrochromicity [85]. For the reduced film, a broad visible absorption band associated with the iron-to-copper CT in cupric ferrocyanide was observed (λ_{max} = 490 nm, ε_{max} = 2 x 10^3 dm^3 mol^{-1} cm^{-1}). This band was absent in the spectrum of the oxidised film, the yellow colour arising from the cyanide to iron CT band at 420 nm for the ferricyanide species.

In a further development, Siperko and Kuwana [86, 88] have produced layered ('sandwich') films of CuHCF and PB. A 'sandwich' electrode with CuHCF (150 Å) as inner layer and PB (500 Å) as outer layer was shown to allow transfer of charge across the film interface, resulting in superposition of the CV responses. In contrast, reversal of the layers produced a distorted CV response. Although visible absorption spectra have not been reported, combination of a second PB analogue with PB in layered structures is likely to allow extension of the polyelectrochromicity of PB itself.

6.5.5 Miscellaneous Metal Hexacyanometallates

Jiang and Zhao [92] have reported the preparation of electrochromic palladium hexacyanoferrate films by the simple immersion (for at least one hour) or potential cycling of conducting substrates (Ir, Pd, Au, Pt, glassy carbon) in a mixed solution of $PdCl_2$ and $K_3Fe(CN)_6$. Resulting modified electrodes gave broad CV responses, assigned to $Fe^{III/II}(CN)_6$, the Pd^{II} sites being electroinactive. Films were orange at > 1.0 V, yellow/green when less positive than (or negative of) 0.2 V.

Indium hexacyanoferrate films [93–97] have been grown by potential cycling in a mixed solution containing $InCl_3$ and $K_3Fe(CN)_6$. The electrodeposition occurs during the negative scans as sparingly soluble deposits of In^{3+} with $Fe(CN)_6{}^{4-}$ were formed [93]. Jin and Dong [97] noted that the resulting films are electrochromic, being white when reduced and yellow when oxidised.

Thin films of cadmium hexacyanoferrate [98], cobalt hexacyanoferrate [99], manganese hexacyanoferrate [100], osmium(IV) hexacyanoruthenate [101], ruthenium hexacyanoferrate [102], silver hexacyanoferrate [8], titanium hexacyanoferrate [103], copper heptacyanonitrosylferrate [104], ferric carbonylpentacyanoferrate [8], ferric pentacyanonitroferrate [8], and Ag(I)–'crosslinked' nickel hexacyanoferrate [105] have all been reported as modified electrodes, although the (likely) electrochromism of these materials has not been commented on.

6.5.6 Mixed Metal Hexacyanoferrates

GC electrodes have been modified with films of mixed metal hexacyanoferrates [100]. CVs of PB–nickel-hexacyanoferrate and PB–manganese-hexacyanoferrate films show electroactivity of both metal hexacyanoferrate components in each mixture. The authors suggest that the mixed metal hexacyanoferrates have a structure where some of the outer sphere iron centres in the PB lattice are replaced by Ni^{2+} or Mn^{2+}, rather than being a codeposited mixture of PB and nickel or manganese hexacyanoferrate. Although film colours are not reported, it seems likely that variation of metal hexacyanoferrate and electrodeposition solution composition would allow colour choice in the resulting poly-electrochromic systems. The approach seems general, with PB–metal hexacyanoferrate (metal = Co, Cu, In, Cr, Ru) modified electrodes also being successfully prepared.

References

[1] Diesbach (1704) cited in Gmelin, Handbuch der Anorganischen Chemie, vol. 59, Teil B.

[2] Mm. Riffault, Vergnaud and Toussaint, ed. M.F. Malepeyre, translated by A.A. Fesquet, 'A Practical Treatise on the Manufacture of Colours for Printing', Sampson Low, Marston, Low and Searle, London, 1874.

[3] K. Fukuda in Pigment Handbook, ed. P.A. Lewis, vol. I, 2nd. ed., Wiley Interscience, New York, 1986, p.357.

[4] Colour Index 3rd. ed., vol. 4, Society of Dyers and Colorists, Bradford, England, 1971, p. 4673.

[5] H. Holtzmann, *Ind. Eng. Chem.*, 37 (1945) 855.

[6] J. Cooper, 'The Man Who Made Husbands Jealous', Bantam Press, Great Britain, 1993.

[7] V.D. Neff, *J. Electrochem. Soc.*, 125 (1978) 886.

[8] K. Itaya, I. Uchida and V.D. Neff, *Acc. Chem. Res.*, 19 (1986) 162.

[9] H.E. Williams, 'Cyanogen Compounds', 2nd. ed., Arnold, London, 1948.

[10] B.M. Chadwick and A.G. Sharpe, *Adv. Inorg. Chem. Radiochem.*, 8 (1966) 83.

[11] A.G. Sharpe, 'The Chemistry of Cyano Complexes of the Transition Metals', Academic Press, New York, 1976.

[12] A.K. Bonnette Jr. and J.F. Allen, *Inorg. Chem.*, 10 (1971) 1613.

[13] B. Mayoh and P. Day, *J.C.S., Dalton Trans.*, (1974) 846.

[14] J.F. Keggin and F.D. Miles, *Nature (London)*, 137 (1936) 577.

[15] H.J. Buser, D. Schwarzenbach, W. Petter and A. Ludi, *Inorg. Chem.*, 16 (1977) 2704.

[16] Y. Yano, N. Kinugasa, H. Yoshida, K. Fujino and H. Kawahara, *Proc. Electrochem. Soc.*, 90–2 (1990) 125.

[17] D. Ellis, M. Eckhoff and V.D. Neff, *J. Phys. Chem.*, 85 (1981) 1225.

[18] K.-C. Ho, *Proc. Electrochem. Soc.*, 94–2 (1994) 170.

[19] R.M.C. Goncalves, H. Kellawi and D.R. Rosseinsky, *J.C.S., Dalton Trans.*, (1983) 991.

[20] R.J. Mortimer and D.R. Rosseinsky, *J. Electroanal. Chem.*, 151 (1983) 133.

[21] R.J. Mortimer and D.R. Rosseinsky, *J.C.S., Dalton Trans.*, (1984) 2059.

[22] K. Itaya, T. Ataka and S. Toshima, *J. Am. Chem. Soc.*, 104 (1982) 4767.

[23] G. Cheng and S. Dong, *Electrochim. Acta*, 32 (1987) 1561.

[24] B.J. Feldman and O.R. Melroy, *J. Electroanal. Chem.*, 234 (1987) 234.

[25] A. Hamnett, S. Higgins, R.J. Mortimer and D.R. Rosseinsky, *J. Electroanal. Chem.*, 255 (1988) 315.

[26] R.J. Mortimer, D.R. Rosseinsky and A. Glidle, *Solar Energy Mater. Solar Cells*, 25 (1992) 211.

[27] K. Itaya and I. Uchida, *Inorg. Chem.*, 25 (1986) 389.

[28] R.J. Emrich, L. Traynor, W. Gambogi and E. Buhks, *J. Vac. Sci. Technol. A*, 5 (1987) 1307.

[29] C.A. Lundgren and R.W. Murray, *Inorg. Chem.*, 27 (1988) 933.

[30] D.J. Beckstead, D.J. De Smet and J.L. Ord, *J. Electrochem. Soc.*, 136 (1989) 1927.

[31] B.J. Feldman and R.W. Murray, *J. Electroanal. Chem.*, 234 (1987) 213.

[32] P.A. Christensen, A. Hamnett and S.J. Higgins, *J.C.S., Dalton Trans.*, (1990) 2233.

[33] D.E. Stilwell, K.W. Park and M.H. Miles, *J. Appl. Electrochem.*, 22 (1992) 325.

[34] K. Itaya, K. Shibayama, H. Akahoshi and S. Toshima, *J. Appl. Phys.*, 53 (1982) 804.

[35] D.W. DeBerry and A. Viehbeck, *J. Electrochem. Soc.*, 130 (1983) 249.

[36] K. Itaya, I. Uchida, S. Toshima and R.M. De La Rue, *J. Electrochem. Soc.*, 131 (1984) 2086.

[37] A. Viehbeck and D.W. DeBerry, *J. Electrochem. Soc.*, 132 (1985) 1369.

[38] K. Honda, J. Ochiai and H. Hayashi, *J.C.S., Chem. Commun.*, (1986) 168.

[39] K. Honda and H. Hayashi, *J. Electrochem. Soc.*, 134 (1987) 1330.

[40] K. Honda and A. Kuwano, *J. Electrochem. Soc.*, 133 (1986) 853.

[41] M.K. Carpenter and R.S. Conell, *J. Electrochem. Soc.*, 137 (1990) 2464.

[42] T. Kase, M. Kawai and M. Ura, SAE Technical Paper Series, 861,362 (1986).

[43] N. Kobayashi, M. Nishikawa, H. Ohno, E. Tsuchida and R. Hirohashi, *J. Soc. Photog. Sci. Technol. Jpn.*, 51 (1988) 375.

[44] K. Honda, M. Fujita, H. Ishida, R. Yamamoto and K. Ohgaki, *J. Electrochem. Soc.*, 135 (1988) 3151.

[45] M.A. Habib, S.P. Maheswari and M.K. Carpenter, *J. Appl. Electrochem.*, 21 (1991) 203.

[46] K.-C. Ho, T.G. Rukavina and C.B. Greenberg, *Proc. Electrochem. Soc.*, 94–2 (1994) 252.

[47] H. Tada, Y. Bito, K. Fujino and H. Kawahara, *Solar Energy Mater.*, 136 (1989) 2131.

[48] H. Tada, Y. Bito, K. Fujino and H. Kawahara, *Proc. Electrochem. Soc.*, 88-23 (1988) 325.

[49] H. Tada, H. Nagayama and H. Kawahara, U.S. Patent No. 4,726,664 (1988).

[50] M.A. Habib and S.P. Maheswari, *J. Electrochem. Soc.*, 139 (1992) 2155.

[51] N. Oyama, T. Ohsaka, M. Menda and H. Ohno, *Denki Kagaku*, 57 (1989) 1172.

[52] T. Miyamoto, M. Ura, S. Kazama, T. Kase and Y. Maeda, U.S. Patent No. 4,645,307 (1987).

[53] H. Inaba, K. Nakase, Y. Yanagida and H. Nishii, U.S. Patent No. 4,773,741 (1988).

[54] M. Kawai, H. Miyagi and M. Ura, U.S. Patent No. 4,801,195 (1989).

[55] T. Kase, T. Miyamoto, T. Yoshimoto, Y. Ohsawa, H. Inaba and K. Nakase, in 'Large-area Chromogenics: Materials and Devices for Transmittance Control', C.M. Lambert and C.G. Granqvist, Editors, pp. 504–517, SPIE Optical Engineering Press, Bellingham, WA (1990).

[56] J.-G. Béraud and D. Deroo, *Solar Energy Mater. Solar Cells*, 31 (1993) 263.

[57] E.A.R. Duek, M.-A. De Paoli and M. Mastragostino, *Adv. Mater.*, 4 (1992) 287.

[58] E.A.R. Duek, M.-A. De Paoli and M. Mastragostino, *Adv. Mater.*, 5 (1993) 650.

[59] M. Morita, *J. Appl. Poly. Sci.*, 52 (1994) 711.

[60] B.P. Jelle, G. Hagen and S. Nodland, *Electrochim. Acta*, 38 (1993) 1497.

[61] B.P. Jelle and G. Hagen, *J. Electrochem. Soc.*, 140 (1993) 3560.

[62] B.P. Jelle and G. Hagen, *Proc. Electrochem. Soc.*, 94–2 (1994) 324.

[63] N. Leventis and Y.C. Chung, *J. Electrochem. Soc.*, 137 (1990) 3321.

[64] N. Kashiwazaki, *Solar Energy Mater. Solar Cells*, 25 (1992) 349.

[65] M.B. Robin, *Inorg. Chem.*, 1 (1962) 337.

[66] H. Inoue and S. Yanagisawa, *J. Inorg. Nucl. Chem.*, 36 (1974) 1409.

[67] R.E. Wilde, S.N. Ghosh and B.J. Marshall, *Inorg. Chem.*, 9 (1970) 2512.

[68] K.P. Rajan and V.D. Neff, *J. Phys. Chem.*, 86 (1982) 4361.

[69] K. Itaya, T. Ataka and S. Toshima, *J. Am. Chem. Soc.*, 104 (1982) 3751.

[70] T. Ataka, T. Sakuhara, M. Sigeno and K. Iwasa, *Japan Display*, (1983) 384.

[71] M.K. Carpenter, R.S. Conell and S.J. Simko, *Inorg. Chem.*, 29 (1990) 845.

[72] D. Shaojun and L. Fengbin, *J. Electroanal. Chem.*, 210 (1986) 31.

[73] S.J. Dong and F.B. Li, *J. Electroanal. Chem.*, 217 (1987) 49.

[74] A.B. Bocarsly and S. Sinha, *J. Electroanal. Chem.*, 137 (1982) 157.

[75] A.B. Bocarsly and S. Sinha, *J. Electroanal. Chem.*, 140 (1982) 167.

[76] A.B. Bocarsly, S.A. Galvin and S. Sinha, *J. Electrochem. Soc.*, 130 (1983) 1319.

[77] S. Sinha, B.D. Humphrey and A.B. Bocarsly, *Inorg. Chem.*, 23 (1984) 203.

[78] B.D. Humphrey, S. Sinha and A.B. Bocarsly, *J. Phys. Chem.*, 88 (1984) 736.

[79] S. Sinha, B.D. Humphrey, E. Fu and A.B. Bocarsly, *J. Electroanal. Chem.*, 162 (1984) 351.

[80] L.J. Amos, M.H. Schmidt, S. Sinha and A.B. Bocarsly, *Langmuir*, 2 (1986) 559.

[81] S. Sinha, L. Amos, M.H. Schmidt and A.B. Bocarsly, *J. Electroanal. Chem.*, 210 (1986) 323.

[82] B.D. Humphrey, S. Sinha and A.B. Bocarsly, *J. Phys. Chem.*, 91 (1987) 586.

[83] L.J. Amos, A. Duggal, E.J. Mirsky, P. Ragonesi, A.B. Bocarsly and P.A. Fitzgerald-Bocarsly, *Anal. Chem.*, 60 (1988) 245.

[84] C. Lin and A.B. Bocarsly, *J. Electroanal. Chem.*, 300 (1991) 325.

[85] L.M. Siperko and T. Kuwana, *J. Electrochem. Soc.*, 130 (1983) 396.

[86] L.M. Siperko and T. Kuwana, *J. Electrochem. Soc.*, 133 (1986) 2439.

[87] L.M. Siperko and T. Kuwana, *Electrochim. Acta*, 32 (1987) 765.

[88] L.M. Siperko and T. Kuwana, *J. Vac. Sci. Technol. A*, 5 (1987) 1303.

[89] P.J. Kulesza and Z. Galus, *J. Electroanal. Chem.*, 267 (1989) 117.

[90] K.N. Thomsen and R.P. Baldwin, *Electroanalysis*, 2 (1990) 263.

[91] G. Horanyi, G. Inzelt and P.J. Kulesza, *Electrochim. Acta*, 35 (1990) 811.

[92] M. Jiang and Z. Zhao, *J. Electroanal. Chem.*, 292 (1990) 281

[93] P.J. Kulesza and M. Faszynska, *J. Electroanal. Chem.*, 252 (1988) 461.

[94] P.J. Kulesza and M. Faszynska, *Electrochim. Acta*, 34 (1989) 1749.

[95] P.J. Kulesza and Z. Galus, *J. Electroanal. Chem.*, 267 (1989) 117.

[96] S. Dong and Z. Jin, *Electrochim. Acta*, 34 (1989) 963.

[97] Z. Jin and S.J. Dong, *Electrochim Acta.*, 35 (1990) 1057.

[98] C.H. Luangdilok, D.J. Arent, A.B. Bocarsly and R. Wood, *Langmuir*, 8 (1992) 650.

[99] J. Joseph, H. Gomathi and G. Prabhakar Rao, *J. Electroanal. Chem.*, 304 (1991) 263.

[100] S. Bharathi, J. Joseph, D. Jeyakumar and G. Prabhakara Rao, *J. Electroanal. Chem.*, 319 (1991) 341.

[101] J.A. Cox and B.K. Das, *J. Electroanal. Chem.*, 233 (1987) 87.

[102] P.J. Kulesza, *J. Electroanal. Chem.*, 220 (1987) 295.

[103] M. Jiang, X.Y. Zhou and Z.F. Zhao, *J. Electroanal. Chem.*, 292 (1990) 289.

[104] Z. Gao, Y. Zhang, M. Tian and Z. Zhao, *J. Electroanal. Chem.*, 358 (1993) 161.

[105] P.J. Kulesza and Z. Galus, *J. Electroanal. Chem.*, 267 (1989) 117.

7 Other Inorganic Systems

The electrodeposition of coloured electrochrome from colourless solution represents electrochromic systems of type 2 (section 2.3.1.2). Reductive deposition of metals in section 7.1 is thus within this category. The other electrochromic materials cited here, being permanently solid, are type 3.

7.1 Deposition of Metals

The deposition of bismuth from acidic solution (0.2 M HCl) follows equation (7.1)

$$\underset{\text{(colourless)}}{2\,Bi^{3+} + 9Br^-} \rightarrow \underset{\text{(black)}}{2Bi^o + 3Br_3^-} \tag{7.1}$$

Copper(II) chloride is also present in solution to mediate the electro-oxidation of the bismuth in electrochromic bleaching. Colouration efficiencies are low [1] at 3 cm^2/C although a contrast ratio *CR* of 50:1 in 30 ms is claimed [2]. The bismuth produced is finely divided, perhaps dendritic, so deposits are not optically reflective. Particulate solid is formed in preference to continuous plate, via an underpotential deposition mechanism.

Electrodeposition of elemental silver has also been used to form coloured images, using either Ag^+ as precursor [3–5] or complexes of silver [6]. The *CR* there for silver-based devices was apparently very high [7].

7.2 Deposition of Colloidal Material

An entirely different system involves the deposition of charged particles to form a thin, compact layer. An exemplar system comprises rod-like particles of TiO_xN_y coated with silica and dispersed in a gel electrolyte, for example, poly(dimethylsiloxane). Application of a strong field to the suspension placed between two ITO electrodes (conducting sides innermost) causes particle alignment and chain formation [8]. This alignment, following the so-called Winslow effect [9], causes light scattering rather than optical absorption and is thus not truly electrochromic. Bleaching of such a display is effected by removing the potential: Brownian motion causes the chains to peptise.

7.3 Intercalation Layers

Pfluger *et al.* [10] have reported an ECD with graphite as a solid-solution intercalation electrode. Many different alkali metal cations may be inserted into graphite sheeting from an aprotic solution, lithium apparently giving the best results. This ECD is poly-electrochromic, switching from brassy black *via* deep blue and light green to golden yellow within the potential range 3–5 V. When the potential was reversed, the ECD reverted back to the brassy black colour, with τ of about 0.2 s. Graphite has also been used as a counter electrode [11].

7.4 Inclusion and Polymeric Systems

Complexes of iron as polyelectrolytes have been used by Oonuk [12], Itaya [13] and Kondo [14], while Oonuk and Itaya used disulphonato bathophenanthroline (bphen) (**I**) complexes of iron :

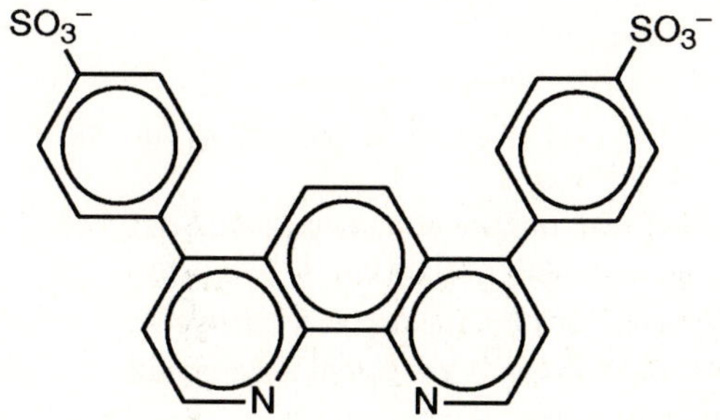

I

Itaya *et al.* [13] and Oyama *et al.* [15] supported the complex on polymeric supports, for example, perfluorocarbon electrolytes, and quote the reaction (7.2)

$$\underset{\text{(red)}}{[Fe(bphen)_3]^{4-}} \rightarrow \underset{\text{(colourless)}}{[Fe(bphen)_3]^{3-}} + e^- \qquad (7.2)$$

In addition to iron, Kondo [14] has used ferrocene bound within phenanthroline ligands to generate electrochromic images.

Electrodes modified with thin electrochromic films have been prepared by Hupp and co-workers [16, 17] using *tris* iron, ruthenium or osmium complexes of pyridine-based ligands, such as 4-methyl-4'-vinylbipyridine (**II**) and 4-(2-pyrrol-1-ylethyl)-2,2'-bipyridine, polymerisation being effected photochemically [16] or electrochemically [17]. All these complexes are coloured if the metal is present in its lower valence state, the iron complex being red, the complexes of ruthenium and osmium being orange and green respectively. Similar polybipyridyl complexes of ruthenium have been prepared by Meyer *et al.*, employing ligand (**II**) [18]. Beer, Mortimer and co-workers have observed that electropolymerised films of $[RuL'(bipy)_2][PF_6]_2$ (L' = 4,4'-*bis*(ferrocenylvinyl)-2,2'-bipyridine, bipy = 2,2'-bipyridine) [19] and $[Ru(L'')_3][PF_6]_2$ (L'' = 4-(3,4-dimethoxystyrl)-4'-methyl-2,2'-bipyridine) [20, 21] complexes exhibit transparent/orange electrochromicity.

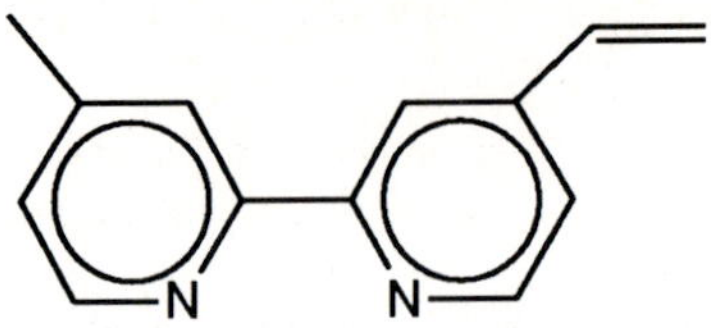

II

The electrochromism of rhodium complexed with 2,2'-bipicoline [22] has also been reported.

7.5 Miscellaneous

Electrochromism has also been reported for many other miscellaneous inorganic systems, as in the following examples.

A series of sintered lead titanate zirconates have been reported [23] of composition $[Pb_{(x-y)}La_y]\ [(Zr_z\,Ti_{(1-z)})_{1-y/4}\ v_{y/4}]O_3$ and $(Pb_{x-3y/2}\ v_{y/2}\ La_y)\ (Zr_z\ Ti_{1-z})\ O_3$ and compositions between the two (v is a cation vacancy; $0 \le x \le 1.01$; $0.01 \le y \le 0.3$ and $0.05 \le z \le 0.95$). Although sintered, the initial state is quite transparent and shows good electrochromic properties.

Nickel-doped strontium titanate $SrTiO_3$ is electrochromic [24]. Amorphous AgRXM (R = S, Se or Te; X = halide and M = P, As or Sb) is electrochromic to some extent [25]. No colour changes are given in either of these two papers.

Bridged ruthenium mixed-valence complexes [26] and thin-film $Pb_{0.875}U_{0.125}F_{2.25}$ [27] are also claimed to be electrochromic. Although not strictly electrochromic, Li_xC_{60} exhibits optical spectra in the near infra red different from C_{60} [28].

The titration indicator ferroin (iron(II) *tris* *o*-phenanthroline) is bright red, going pale on oxidation [29]. It is thus a candidate for electrochromism [30], as indeed would be any redox indicator, complexed or not. Furthermore, any coloured metal-ion complex susceptible to a redox change will in general undergo an accompanying colour change, and will therefore be electrochromic.

References

[1] J.P. Ziegler and B.M. Howard, *Proc. Electrochem. Soc.*, 94-2 (1994) 158.

[2] B.M. Howard, J.P. Ziegler, M. McKee, N. Tornberg and K. Caudy, *Proc. Electrochem. Soc.*, 93–26 (1993) 353.

[3] S. Zaromb, *J. Electrochem. Soc.*, 109 (1962) 903.

[4] S. Zaromb, *J. Electrochem. Soc.*, 109 (1962) 912.

[5] J. Mantell and S. Zaromb, *J. Electrochem. Soc.*, 109 (1962) 992.

[6] I. Camlibel, S. Singh, H.J. Stocker, L.G. van Ultert and G.J. Zydzik, *Appl. Phys. Lett.*, 33 (1978) 793.

[7] H.J. Byker, *Proc. Electrochem. Soc.*, 94–2 (1994) 3.

[8] Y. Saito, M. Hirata, H. Tada, M. Hyodo and H. Kawahara, *Proc. Electrochem. Soc.* 94–2 (1994) 354.

[9] W.M. Winslow, *J. Appl. Phys.*, 20 (1949) 1137.

[10] P. Pfluger, H.U. Künzi and H.J.Güntherodt, *Appl. Phys. Lett.*, 35 (1979) 771.

[11] K. Kuwabara and Y. Noda, *Solid State Ionics*, 61 (1993) 303.

[12] Y. Oonuk and A. Kondo, *Jpn. Kokai. Tokkyo Koho.* JP 62,104,891, cited in *Chem. Abstr.* 107: P187,533*u*.

[13] K. Itaya, H. Akahoshi and S. Toshima, *J. Electrochem. Soc.*, 129 (1982) 762.

[14] Y. Kondo, A. Nishino, T. Yamamoto and Y. Osada, *Jpn. Kokai. Tokkyo Koho.* JP 62,621,825 cited in *Chem. Abstr.* 107: P155,849*k*.

[15] N. Oyama, M. Kitagawa, M. Iwaku, H. Inaba and T. Tatsuma, Extended Abstracts, 183rd Meeting of the Electrochemical Society, Hawaii, 1993. Abstract 1686.

[16] P. Subramanian, H.-T. Zhang and J.T. Hupp, *Inorg. Chem.*, 31 (1992) 1540.

[17] H.-T. Zhang, P. Subramanian, O. Fussa-Rydel, J.C. Bebel and J.T. Hupp, *Solar Energy Mater. Solar Cells*, 25 (1992) 315.

[18] R.M. Leasure, W. Ou, J.A. Moss, R.W. Linton and T.J. Meyer, *Proc. Electrochem. Soc.*, 94–2 (1994) 222.

[19] P.D. Beer, O. Kocian and R.J. Mortimer, *J.C.S., Dalton Trans.*, (1990) 3283.

[20] P.D. Beer, O. Kocian, R.J. Mortimer and C. Ridgway, *J.C.S., Faraday Trans.*, 89 (1993) 333.

[21] P.D. Beer, O. Kocian, R.J. Mortimer and C. Ridgway, *J.C.S., Dalton Trans.*, (1993) 2629.

[22] N. Kobayashi, M. Norishisa, N. Mineta and R. Hirohashi, *Nippon Shashin Gakkaishi*, 53 (1990) 389, cited in *Chem. Abstr.* 115 (8) 80,873*t*.

[23] T. Takagi, K. Ametani and K. Shimizu, *Jpn. Kokai Tokkyo Joho*, JP 62,182,159 (1987), cited in *Chem. Abstr.* 107: P241,455*k*.

[24] K. Sarat, *J. Appl. Phys.*, 50 (1979) 5001.

[25] J. Portier, J.M. Reau, H.W. Sun, B. Tanguy, R. Astier, C. Combie, M. Maurin, A. Pradel and M. Ribes, French Patent: FR 2,594,116 (1987) cited in *Chem. Abstr.* 107: P248,339*j*.

[26] D.H. Oh and S.G. Boxer, *J. Am. Chem. Soc.*, 112 (1990) 8161.

[27] M. Addou and A. Kkadiri, *J. de Chimie Physique.*, 89 (1992) 1477.

[28] J.D. Klein, A. Yen, R.D. Rauh and S.L. Clauson, *Appl. Phys. Lett.*, 63 (1993) 599.

[29] E.B. Sandell, 'Colorimetric Determination of Traces of Metal', 3rd Edn., Interscience, New York, 1959. p 524.

[30] S.S. Zhang, X.P. Qui, W.H. Chou, Q.G. Liu, L.L. Lang and B.Q. Xing, *Solid State Ionics*, 52 (1992) 287.

B Organic Systems

8 Bipyridilium Systems

8.1 Introduction

The next major group of electrochromes are the bipyridilium species formed by the diquaternising of 4,4'-bipyridyl to form 1,1'-disubstituted-4,4'-bipyridilium salts (Fig. 8.1). The positive charge shown localised on N is in general delocalised over the rings. The compounds are formally named as 1,1'-di-substituent-4,4'-bipyridilium if the two substituents at nitrogen are the same, and as 1-substituent-1'-substituent'-4,4'-bipyridilium should they differ. The anion X^- in Fig. 8.1 need not be monovalent and can be part of a polymer. The molecules are zwitterions when the substituents at nitrogen bear a negative charge [1, 2].

A convenient abbreviation for any bipyridyl unit regardless of its redox state is 'bipm' with its charge indicated. The literature of these compounds contains several trivial names. The most common name for these salts is 'viologen' following Michaelis [3, 4], who noted the violet colour formed when 1,1'-dimethyl-4,4'-bipyridilium undergoes a one-electron reduction to form a radical cation. 1,1'-Dimethyl-4,4'-bipyridilium is therefore called methyl viologen (MV) in this nomenclature. Another extensively used name is 'paraquat', PQ, after the ICI brand name for the widely used herbicide, 1,1'-dimethyl-4,4'-bipyridilium. Other bipyridilium species than the dimethyl are called *substituent* paraquat.

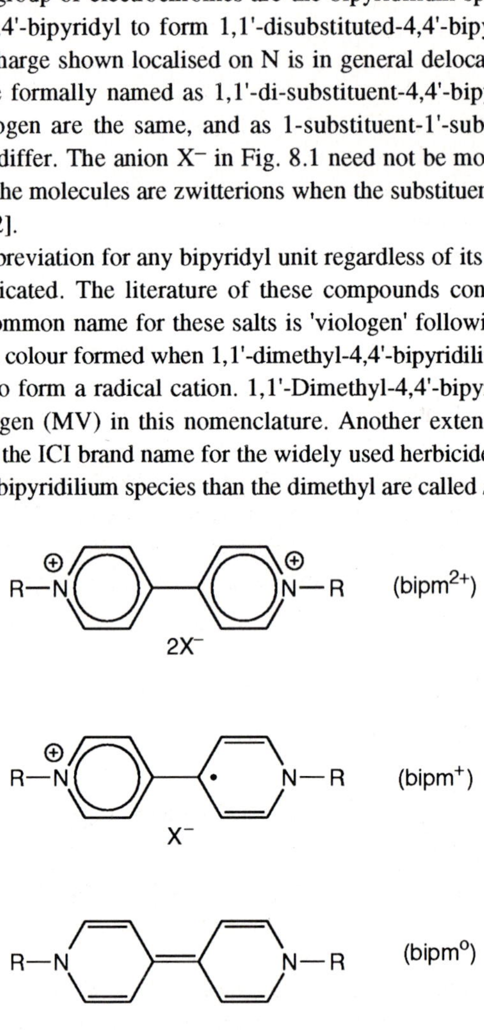

Fig. 8.1 The three common bipyridyl redox states. Different substituents as R_1 and R_2 may be attached to form unsymmetrical species.

There are several excellent reviews of this field: 'The Bipyridines' (1984) by Summers [5] deals at length with syntheses and properties of 4,4'-bipyridine; 'Cation Radicals' (1976) by Bard, Ledwith and Shine [6], has a section on bipyridilium radical cations; 'The Electrochemistry of the Viologens' (1981) by Bird and Kuhn [7] is particularly relevant to this chapter.

8.2 Bipyridilium Redox Chemistry

There are three common bipyridilium redox states: a dication ($bipm^{2+}$), a radical cation ($bipm^{+\bullet}$) and a di-reduced neutral compound ($bipm^{o}$). The dication is the most stable of the three and is the species purchased or first prepared in the laboratory. It is colourless when pure unless optical charge transfer with the counter anion occurs. Such absorbances are feeble for anions like chloride but are stronger for CT-interactive anions like iodide [8]; $PQ^{2+}\ 2I^{-}$ is brilliant scarlet.

Reductive electron transfer to the dication forms a radical cation

$$\underset{\text{(colourless)}}{bipm^{2+} + e^{-}} \quad \rightarrow \quad \underset{\text{(intense colour)}}{bipm^{+\bullet}} \qquad (8.1)$$

Bipyridilium radical cations are amongst the most stable organic radicals, and may be prepared as air-stable solids [9, 10]. In solution the colour of the radical will persist almost indefinitely in the absence of oxidising agents like ferricyanide [11] or periodate, although its reaction with molecular oxygen is particularly rapid [12]. The stability of the radical cation is attributable to the delocalisation of the radical electron throughout the π-framework of the bipyridyl nucleus, the 1 and 1' substituents commonly bearing some of the charge.

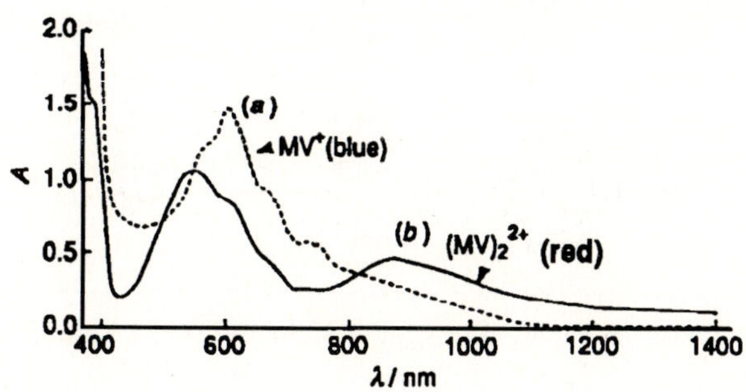

Fig. 8.2 UV-vis spectra of the methyl viologen radical cation in aqueous solution. (a) Monomeric (blue) radical cation and (b) red radical-cation dimer, the sample also containing a trace of monomer. (Figure reproduced from ref. [66] by permission of the Royal Society of Chemistry.)

Electrochromism occurs in bipyridilium species because, in contrast to the bipyridilium dications, radical cations are intensely coloured owing to optical charge transfer between the (formally) +1 and zero valent nitrogens, in a simplified view of the phenomenon (in fact, because of the delocalisation referred to, the source of the colour is probably better viewed as an internal photo-effected electronic excitation). The colours of radical cations depend on the substituents on the nitrogen. Simple alkyl groups, for example, promote a blue/violet colour whereas aryl groups generally impart a green hue to the radical cation. Manipulation of the substituents at N to attain the appropriate molecular orbital energy levels can, in principle, tailor the colour as desired. The colour will in general depend on solvent*.

The molar absorptivity (or extinction coefficient) ε for the methyl viologen radical cation is large; for example, in water ε = 13,700 $dm^3\ mol^{-1}\ cm^{-1}$ when extrapolated to zero concentration [14]. ε is usually somewhat solvent dependent [15]. A few values of wavelength maxima and ε are listed in Table 8.1. The data refer to monomeric radical-cation species unless stated otherwise.

Table 8.1 Optical Data for Some Bipyridilium Radical Cations.

R	anion	solvent	λ_{max}/nm	$\varepsilon/dm^3\ mol^{-1}\ cm^{-1}$	ref.
Methyl	Cl^-	H_2O	605	13,700	[15]
Methyl	I^-	H_2O/MeCN	605^a	10,060	[20, 21]
Methyl	Cl^-	H_2O	606	13,700	[14]
Methyl	Cl^-	MeCN	607	13,900	[15]
Methyl	Cl^-	MeOH	609	13,800	[15]
Methyl	Cl^-	EtOH	611	13,800	[15]
Methyl	Cl^-	H_2O	604	16,900	[22]
Ethyl	ClO_4^-	DMF	603	12,200	[23]
Heptyl	Br^-	H_2O	$545^{b,c}$	26,000	[24]
Octyl	Br^-	H_2O	543c	28,900	[25]
Benzyl	Cl^-	H_2O	604	17,200	[26]
p-CN-Ph	BF_4^-	PC	674	83,300	[27]
p-CN-Ph	Cl^-	H_2O	$535^{b,c}$	—	[28]

Key: [a] Estimated from reported spectra. [b] Solid on OTE. [c] Solution-phase radical-cation dimer.

* Kosower's solvent *Z* values (optical CT energies) in reference [13] were determined using the different but related system comprising 4-carboethoxy-1-methylpyridinium iodide.

Comparatively little is known about the third redox form of the bipyridilium series, the di-reduced or so-called 'di-hydro' [16] compounds formed by one-electron reduction of the respective radical cation (reaction 8.2)

$$\underset{\text{(intense colour)}}{\text{bipm}^{+\bullet} + \text{e}^-} \rightarrow \underset{\text{(weak colour)}}{\text{bipm}^{\text{o}}} \qquad (8.2)$$

This product may also be formed by direct two-electron reduction of the dication.

Di-reduced compounds are sometimes called 'bi-radicals' [17] because of their reactivity, although magnetic susceptibility measurements have shown such species to be diamagnetic [18] in the solid state, indicating that spins are paired. In fact, di-reduced bipm^{o} compounds are simply reactive amines [19]. The intensity of the colour exhibited by bipm^{o} species is low since no optical charge transfer or internal transition corresponding to visible wavelengths is accessible.

8.3 Bipyridilium Species for Inclusion Within ECDs

The most extensive literature on a bipyridilium compound is that for 1,1'-dimethyl-4,4'-bipyridilium, also called 'methyl viologen' (MV). The write–erase efficiency of an ECD using aqueous MV as the electrochrome would be low since the MV electrochrome is very soluble in water in both its dicationic and radical-cation states. The write–erase efficiency of MV-based ECDs may be improved by retarding the rate at which the radical-cation product of electron transfer diffuses away from the electrode and in to the solution bulk either by tethering the dication to the surface of an electrode (a 'derivatised' electrode), or by immobilising the viologen species within a semi-solid electrolyte. These approaches, with the methyl viologen behaving as a pseudo permanent-solid electrochrome, are described in sections 8.3.1 and 8.3.2 respectively.

The solubility/diffusion problem can be avoided by the use of viologens having long alkyl-chain substituents at nitrogen, for which the coloured radical-cation product of reaction (8.1) is insoluble, so here the viologen is a solution-to-solid electrochrome, as discussed in section 8.3.3.

8.3.1 Derivatised Electrodes for ECD Inclusion

Wrighton *et al.* [29–31] have often derivatised electrodes with bipyridilium species, initially using substituents at N consisting of a short alkyl chain terminating in the trimethoxysilyl group, which may bond to the oxide lattice on the surface of an OTE. With chemical tethering of this type, Wrighton and Bookbinder [29] bonded the viologen (**I**) and benzyl viologen [30] derivatives to electrode surfaces.

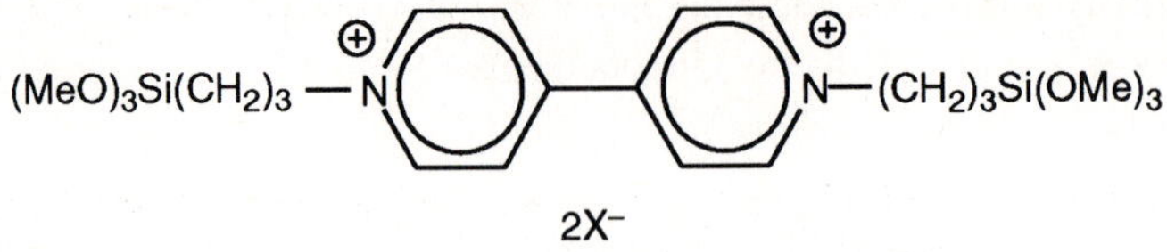

I

More recently, Wrighton *et al.* diquaternised a bipyridilium nucleus with a short alkyl chain terminating in pyrrole (which was bonded to the alkyl chain at nitrogen [31]) – see (**II**); anodic polymerisation of the pyrrole allowed an adherent film of polypyrrole to form on the electrode surface [31], thus attaching the bipyridilium units to the electrode.

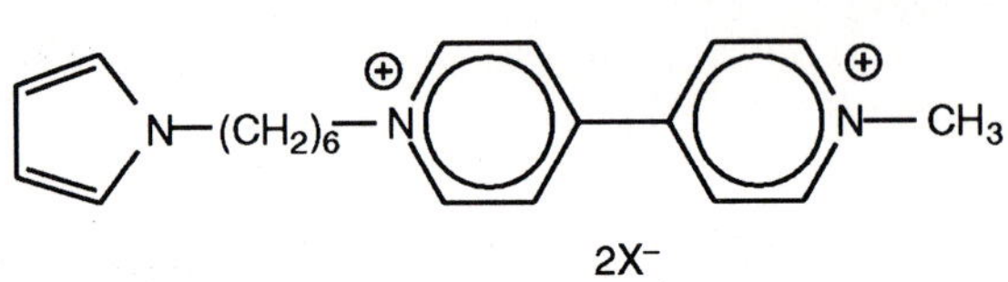

II

Itaya *et al.* [32] used polymeric electrolytes, but with the salt electrostatically bonded to a poly(styrene sulphonate) electrolyte. A bipyridilium salt of poly(*p*- or *m*-xylyl)-4,4'-bipyridilium bromide (**III**, shown in the *p* form) was the electrochrome. The interaction between the cationic bipyridilium nucleus and the sulphonyl group is coulombic. The electrode was prepared by dipping the conducting substrate into solutions of electrochrome-containing polymer which, after drying, is insoluble in aqueous solution. Polymeric bipyridilium salts have also been prepared by Simon and Moore [33], Moutet *et al.* [34], Berlin *et al.* [35], Stile [36], Factor and Heisolm [37] Sato and Tamamura [38] and Willman and Murray [39].

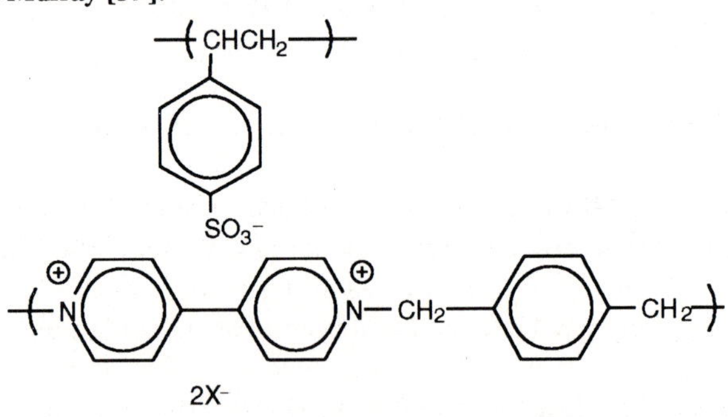

III

8.3.2 Immobilised Bipyridilium Electrochromes for ECD Inclusion

A different method of ensuring a high write–erase efficiency is to embed the bipyridilium salt within a polymeric electrolyte; for example, Sammells and Pujaru [40, 41] suspended heptyl viologen (HV) in polyAMPS, while Calvert *et al.* [42] used methyl viologen also in polyAMPS. Both groups report an excellent long-term write–erase efficiency and good electrochromic memory.

8.3.3 Soluble-to-Insoluble Bipyridilium Electrochromes for ECD Inclusion

8.3.3.1 Devices

The bipyridilium species most thoroughly studied for electrochromic applications is 1,1'-diheptyl-4,4'-bipyridilium (heptyl viologen, HV) as the dibromide salt. HV^{2+} dication is soluble in water, but forms an insoluble film of crimson-coloured radical-cation salt adhering strongly to the electrode surface following a one-electron reduction, as in reaction (8.1).

The first ECD using bipyridilium salts was reported by Schoot *et al.* [43] (of the Philips Laboratories) in 1973. Philips had submitted Dutch patents in 1970 [44] for HV as the electrochrome, while in 1971 ICI patented the aryl-substituted viologen 1,1'-*bis*(*p*-cyanophenyl)-4,4'-bipyridilium ('cyanophenyl paraquat' or 'CPQ') [45].

Schoot's ECD device had a contrast ratio of 20 : 1, an erase time of 10 to 50 ms [43], and cycle life of more than 10^5 cycles. In the Philips ECD, heptyl viologen was used because reduction of the dication formed a durable film on the electrode, whereas shorter alkyl chains yield radical-cation salts which are slightly soluble.

8.3.3.2 The Effect of the Electrode Substrate

Van Dam and Ponjée [46] examined the effect that variations in the length of the alkyl chain have on the film-forming properties of the radical cation as the bromide salt (table 8.2), and redox potentials have been added to this table from ref [7]. The heptyl chain produces the first truly insoluble viologen radical-cation salt as the length of the alkyl chain is increased.

From table 8.2 it is clear that an effective chain length in excess of 4 CH_2 units is necessary for stable films to form. The radical-cation salt of cyanophenyl paraquat (CPQ) is more insoluble in water than is $HV^{+\bullet}$, yet the dicationic salt is very soluble. The solubility product K_{SP} of $HV^{+\bullet}$ Br^- in water is [46] 3.9×10^{-7} M^2.

The radical cations of viologen species containing short alkyl chains have a blue colour becoming blue-purple in concentrated solution [21]. The colour of the radical cation tends to crimson as the length of the alkyl chain increases, largely owing to increasing incidence of radical cation dimerisation; the dimer of alkyl-substituted radical cations is red [21].

By comparison, aryl substituted viologens generally form green or dark red radical cations salts. Also, dication solubility and radical cation stability (in thin films) are both improved by using aryl substituents. ICI used the aryl-substituted viologens (*p*-cyanophenyl or CPQ) in their ECD since the electrochromic colour of the heptyl viologen radical cation was considered to show insufficient intensity: the molar absorptivity (and therefore the *CR*) of aryl-substituted viologens is greater than that of alkyl-substituted viologens (see Table 8.1). The radical cation of CPQ apparently [45] has a stability superior to other aryl viologen radical cations.

Table 8.2 Symmetrical Viologens: The Effect of Varying the Alkyl Chain Length on Radical Cation Film Stability (refs. [46] and [7]). $E_1{}^o$ values are quoted *vs* the SCE and refer to viologen salts with the parenthesised anion.

Substituent	Effective Length of R (units of CH_2)	Solid Bromide Salt Film on Pt?	Colour	$E_1{}^o$/mV
Methyl	1	No	blue	−688 (Cl^-)
Ethyl	2	No	blue	−691 (Cl^-)
Propyl	3	No	blue	−690 (Br^-)
Butyl	4	No	blue	−686 (Br^-)
Pentyl	5	Yes	purple	−686 (Br^-)
Hexyl	6	Yes	purple	−710 (Br^-)
Heptyl	7	Yes	mauve	−600 (Br^-)
Octyl	8	Yes	crimson	−705 (Br^-)
iso-pentyl	4	Yes	purple	−696 (Br^-)
Benzyl	4–5	Yes	mauve	−573 (Cl^-)
$CH_3(Cl)CH_2OCH_2$–	4	No	—	
CH_3-CH=CH–CH_2–	4	No	—	
H-CH=CH–$(CH_2)_3$–	4–5	No	—	
NC–C_3H_6–	4–5	No	—	−362[a] (Cl^-)

[a] polarographic $E_{1/2}$ value.

8.3.3.3 The Effect of the Counter Anion

The counter anion in the viologen salt may crucially affect the ECD performance. Different counter ions yield solid radical-cation products of electrodeposition having a wide range of solubilities and chemical stabilities [47]. For example, $CPQ^{+\bullet}$ is oxidised chemically by the nitrate ion [47] via a rapid but complicated mechanism [47]. Studies of counter-ion effects may be performed using cyclic voltammetry (e.g. ref. [48]) or by observing the time dependence of an ESR trace (which is proportional to $[bipm^{+\bullet}]$) [47]. ICI used CPQ^{2+} SO_4^{2-} in their ECDs [45].

The properties of heptyl viologen radical-cation films are also anion dependent: van Dam and Ponjée [46] have investigated the effect of the anion, as has Jasinski [48] (of Texas Instruments), who found the optimum anion in water to be dihydrogen phosphate. Anions found to be compatible with ECD operation were dihydrogen phosphate, sulphate, fluoride, formate and acetate. Bromide, chloride, tetrafluoroborate and perchlorate were also found to be satisfactory (which was also the conclusion of van Dam and Ponjée [46]). Heptyl viologen salts of bicarbonate (at pH 5.5), thiocyanate, tetrahydroborate, hexafluorophosphate, tetrafluoroantimonate and tetrafluoroarsenate are all water insoluble. In addition to $CPQ^{+\bullet}$, $HV^{+\bullet}$ is also oxidised by the nitrate ion [48] presumably following a similar reaction mechanism.

Jasinski [48] has quoted reduction potentials for aqueous HV^{2+} determined using various metals as electrode substrates, and with a variety of anions (table 8.3). Many other redox potentials for mono-reduction of bipyridilium salts are quoted in the review by Bird and Kuhn [7].

8.3.3.4 Kinetics and Mechanism

Bruinink and van Zanten [49] (Philips Laboratories) studied the kinetics of HV^{2+} dibromide reduction in response to a potential step; similarly Jasinski [50] used an OTE, both groups finding the kinetics of mono-reduction to depend on the electrode history and mode of preparation. For HV^{2+}–$(H_2PO_4^-)_2$, the data obtained do not allow any distinction between two possible but different reduction mechanisms. Jasinski quotes a two-stage process (also favoured by van Dam [46] and Schoot *et al.* [43]) in which solution-phase HV^{2+} is reduced to form the radical cation (reaction 8.1), followed by ion-pairing and precipitation of the salt

$$HV^{+\bullet} + \text{Anion}^- \rightarrow HV^{+\bullet}\,\text{Anion}^-(s) \qquad (8.3)$$

The second mechanism has the two steps occurring simultaneously.

Bruinink and Kregting [51] (*cf.* Jasinski [53]), while citing the two-step mechanism, found the reduction process to be compatible also with a theoretical model of metal deposition derived by Berzins and Delahay [52].

Barradas *et al.* [54] (reducing HV^{2+} at an OTE disc) found the reduction process proceeded via a nucleation step. At low overpotentials of mono-reduction, the rate of reduction was controlled by electron transfer, and at high overpotentials, the nucleation process, once initiated, was sufficiently fast that the crystal-growth process was controlled by mass transport. Hemispherical diffusion was inferred, creating diffusion zones which could overlap, after formation, and lead to semi-infinite planar diffusion. In summary, the process may be written as electron transfer → nucleation → hemispherical diffusion → linear diffusion. Barradas found the process too complex to allow precise mathematical models of deposition to be used.

The morphology of $HV^{+\bullet}$ films has been addressed by Barna [55] (Texas Instruments Laboratories). Deposited films are partially crystalline but largely amorphous, acquiring a greater degree of crystallinity with time; this acquisition of crystallinity is probably the cause of additional sharp peaks observed during cyclic voltammetry of heptyl viologen films [7, 46, 48, 50]. The time-dependent change within deposits of $HV^{+\bullet}$ on an OTE has been observed by Goddard *et al.* [56] using UV-vis spectroscopy with a novel cycling technique (now rendered obsolete by the advent of diode-array spectrophotometry).

Table 8.3 The Effect of Supporting Electrolyte Anion, and of Electrode Substrate, on the Reduction Potentials[a] of Heptyl Viologen (ref. [48]). E_{pc} are cited *vs* the SCE.

Anion	$E_{pc}(1)$/V	$E_{pc}(2)$/V on Au	$E_{pc}(1)$/V	$E_{pc}(2)$/V on Pt	$E_{pc}(1)$/V	$Epc(2)$/V on Ag
Bromide (0.3 M)	–0.698	–1.008	–0.708	(< –0.818)	–0.708	–0.978
$H_2PO_4^{2-}$ (2 M)	–0.668	–1.048	–0.668	(< –0.818)	–0.668	
Formate (0.4 M)	–0.848	–0.928			–0.828	–0.948
HCO_3^{2-} (1 M)	–0.768	–0.958			–0.778	–0.948
Acetate (0.5 M)	–0.828	–0.928				
Fluoride (1 M)[b]	–0.818[c]	–0.878[c]		–0.868		–0.808
Sulphate (0.3 M)	–0.818[c]	–0.928[c]	–0.848[b]	–0.898	–0.798	–0.918

Key: [a] Reduction potentials determined at pH 5.5. [b]Millimolar bipyridilium dication employed for measurement. [c] No colour.

Bewick *et al.* [57, 58], using diode-array optical spectroscopy, have investigated HV^{2+} dibromide and many asymmetric bipyridilium salts (that is, with substituents at N and N' being different). The initial solid product of mono-reduction was considered to be $HV^{+\bullet}$ radical cation in a salt which incorporates HV^{2+} dication [58]. Subsequent aging effects and previously inexplicable additional CV peaks are explained in terms of such a composite form of solid deposit. A similar explanation for the complicated cyclic voltammetric behaviour observed during the formation of solid $CPQ^{+\bullet}$ radical cation salt has also been advanced recently [9, 59].

Aqueous solutions of heptyl viologen dication are micellar [60]. Electrochemistry at these micelles is envisaged to proceed in discrete steps, dication on the micelle periphery being reduced preferentially. Association of bipyridilium species to form π-dimers is a well documented phenomenon [20, 21] for the radical cation, but is not so well attested for the dication, although see references [9] and [60–62]. Barclay [60] (of IBM) quotes a critical micelle concentration for the HV^{2+} dication of 10^{-2} M in aqueous bromide solution.

Scharifker and Wehrmann [63] investigated phase changes within radical-cation salt deposits of $HV^{+\bullet}$ and benzyl viologen radical cation. Golden and Przyluski [64] found the aging effect to be due, in part, to the dimerisation of radical cation in solution and, with Belinko [17], suggest that device failure is also due to production of di-reduced bipyridilium ($bipm^{o}$) as a minor electrode product. Recent work has shown that the radical cation dimer is electrochemically only quasi-reversible, that is, slow electro-oxidation occurs [65], hence the observed failure of devices containing traces of dimer. The formation of diamagnetic HV^{o} (at large negative potentials) should be avoided since it also reacts electrochemically quasi-reversibly in aqueous solution [7].

Belinko [17] investigated the write–erase efficiency of $HV^{+\bullet}$ films by cyclic voltammetry, making the lower scanning limit progressively more negative to deliberately generate $bipm^{o}$. When $bipm^{o}$ is formed, reaction with $bipm^{2+}$ forms radical-cation dimer (which is electrochemically quasi-reversible) as the initial product [65, 66]

$$bipm^{2+} + bipm^{o} \rightarrow (bipm^{+\bullet})_2 \rightarrow 2\ bipm^{+\bullet} \qquad (8.4)$$

in the so-called 'comproportionation' reaction. Subsequent dimer dissociation yields monomeric radical cation [66]. In general, solid deposits of $bipm^{+\bullet}$ exhibit spectroscopic IR bands attributable to $(bipm^{+\bullet})_2$ [67].

Comproportionation between dication $bipm^{2+}$ and $bipm^{o}$, forming radical cation, occurs in solution and also in the solid state, as observed for CPQ^{o} and CPQ^{2+} from aqueous electrolyte in the presence of ferrocyanide ion [47, 68].

Films of $HV^{+\bullet}$ salt have been studied by many techniques including UV-vis spectro-electrochemistry [58, 69–71], ESR [72], Raman spectroscopy [73–75], photoacoustic spectroscopy [76, 77], photothermal spectroscopy [78] and, most recently, by use of the quartz-crystal microbalance [79].

Benzyl viologen has also been extensively investigated since it will also form an insoluble film of radical-cation salt following one-electron reduction [56, 63, 70, 80]. Thus, the colour changes on a Pt electrode have been monitored by diffuse reflectance [80] (Figs. 8.3 a and b).

(a)

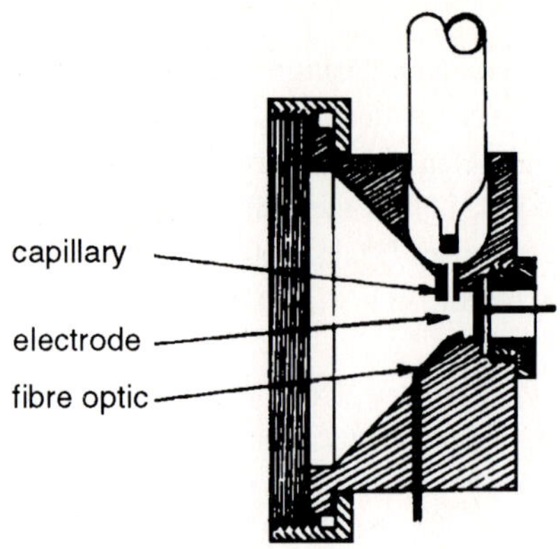

(b)

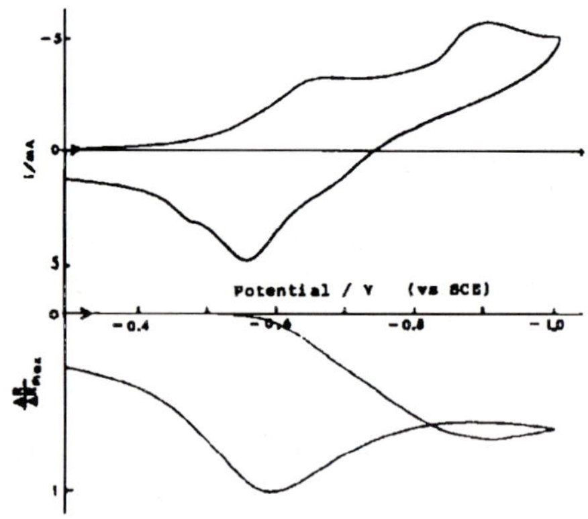

Fig. 8.3 Reflectance voltammetry (a) Black nylon cell: top tube with narrowed sintered end contains SCE, and a Pt wire circle (not shown) about window is the counter electrode. White or monochromated illumination through window is reflected, with absorption, into the optical fibre, which is attached to a photodiode. The photocurrent generated is proportional to the reflected intensity. (b) Two reductive peaks of benzyl viologen BV^{2+} are accompanied by reflectance changes (decreasing on $BV^{+\bullet}$ formation, levelling off as the less-intensely coloured BV^{0} is formed). The reversal, slightly more complicated, is discussed in ref. [80] (Reproduced from D.R. Rosseinsky, J.D. Slocombe, A. Soutar, P.M.S. Monk, and A. Glidle, J. Electroanal. Chem., 259 (1989) 233, copyright (1989), with permission from Elsevier Science Ltd., The Boulevard, Langford Lane, Kidlington, OX5 1GB, UK.)

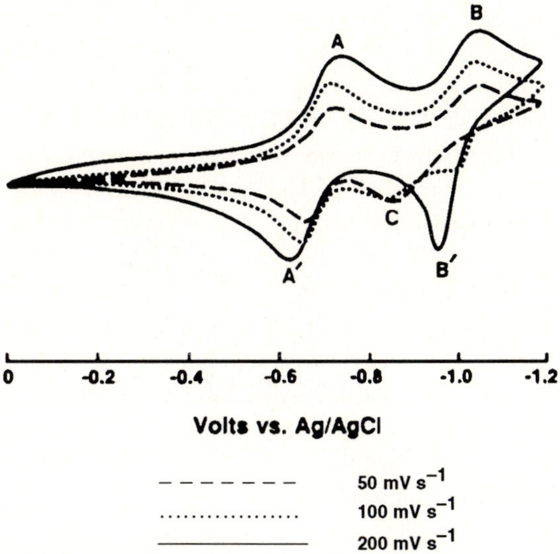

Fig. 8.4 *Cyclic voltammogram on glassy carbon of aqueous methyl viologen dichloride (1 mM) in KCl supporting electrolyte (0.1 M). Note the evidence of reaction 8.4: the oxidation peak for spin-paired radical cation dimer (C) is prominent while the peak for re-oxidation of bipm^{o} (B') is greatly diminished at slow scan rates. (Figure reproduced from ref. [81] with permission of the Society of Applied Spectroscopy.)*

8.3.3.5 The Write–erase Efficiency

The mechanism of deposition is discussed here at length since the nature of the solid deposit is important. A fresh film of $HV^{+\bullet}$ is amorphous, even [48] and smooth [79], yet soon after deposition (< 10 s [56]) the film appears patchy as the aging process occurs, which probably involves ordering (crystallisation) of radical moieties. Re-oxidation of the film to bleach the colour is rapid for fresh $HV^{+\bullet}$ films, but patchy films that show signs of aging are more difficult to oxidise, requiring a higher potential or a longer re-oxidation time.

Secondly, after prolonged cycling between the coloured and bleached states, bipyridilium ECD devices form an unsightly yellow-brown stain on the electrode. There is some evidence that this stain is a form of crystalline radical-cation salt [47] containing spin-paired radical-cation dimer.

The first method used to prevent the non-erasure of films of $HV^{+\bullet}$ salt was to add an 'auxiliary redox couple' (that is, an electron mediator) to the dication containing electrolyte solution; mediators used have included hydroquinones [44], ferrous ion [45], ferrocyanide [45, 73, 82], or cerous ion [40, 41]. During electrochromic reduction, $bipm^{2+}$ is reduced to $bipm^{+\bullet}$ as usual but, during re-oxidation at a positive potential, it is the mediator (e.g. ferrocyanide) that is oxidised electrochemically. The oxidised form of the mediator (here, ferricyanide) allows for chemical oxidation of the radical cation film, to re-form dication. For aryl viologens in aqueous solution, a mediator is always necessary to ensure complete colour removal on re-oxidation [27]. Ferrocyanide is known to form a charge-transfer complex with methyl viologen dication [83, 84] and also with the dications of HV [82] and CPQ [27, 28, 85]. The unsightly yellow-brown stains still persist, however, even with the HV^{2+} and CPQ^{2+} systems containing $K_4Fe(CN)_6$ [45, 73]. In a notable advance, addition of β-cyclodextrin to the voltammetry solution has been found to impede the formation of yellow-brown stains [82], probably by encapsulation of the dication within the cavity of the the cyclodextrin in a guest-host relationship. Because close contact between bipyridilium dications is difficult in such a guest-host relationship, association of $bipm^{2+}$ in solution [60] is thereby prevented, and alignment of $bipm^{+\bullet}$ species in the solid deposit is impossible.

Other attempts to stop the aging phenomenon have used different 'modified' bipyridilium compounds [86–88]. For example, Bruinink *et al.* [86] prepared the compound (**IV**) in which the two pyridinium rings are separated by methylene linkages.

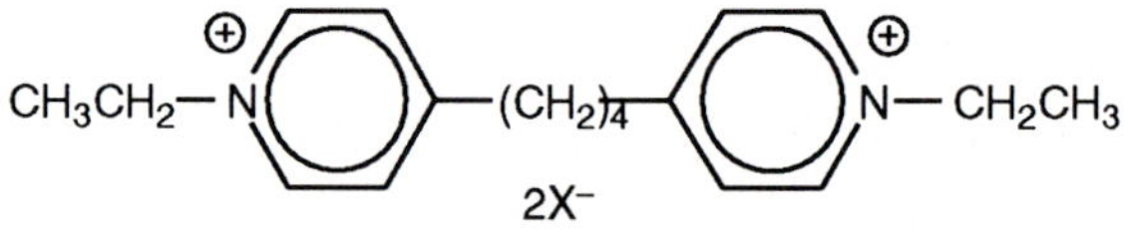

IV

To a similar end, Barna and Fish [87] prepared asymmetric bipyridilium salts, that is species in which $R_1 \neq R_2$ (Fig 8.1), thereby inhibiting the crystallisation process: for example, they made a compound with $R_1 = C_7H_{15}$ and $R_2 = C_{18}H_{37}$. Barltrop and Jackson [88] have also prepared such asymmetric viologens, and a quaternised 3,8-phenanthroline salt (**V**), together with a series of nuclear-substituted bipyridyls (that is, species in which substituents are directly bonded to carbon in the pyridine rings). Again, films with superior write–erase properties were formed.

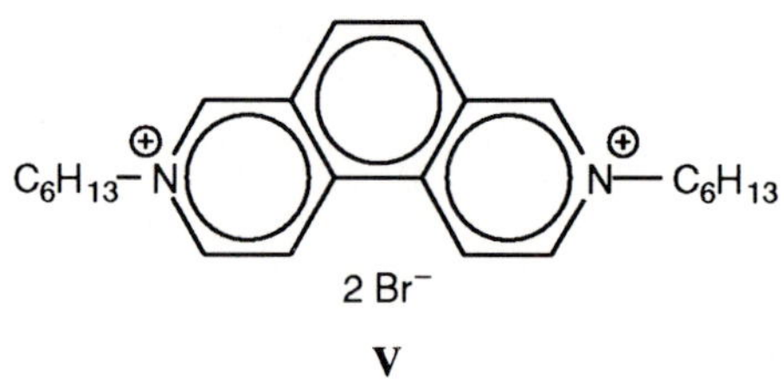

V

Despite the drawbacks, many prototype bipyridilium ECD devices have been made [44, 45, 60, 89, 90]. An impressive device from the IBM laboratories utilised a 64 × 64 pixel integrated ECD device with 8 levels of grey tone of heptyl viologen [90] on a 1 inch square silicon chip, to give quite detailed images (Fig. 8.5). These have not been exploited further owing to LCD competition, though they may still have a size advantage in large devices.

Fig. 8.5 Reproduction of an IBM electrochromic image displayed on a 64 × 64 pixel integrated ECD device with 8 levels of grey tone of heptyl viologen. (Taken from ref. [90] by permission of the copyright holder Dr E.R. Howells.)

8.4 Recent Developments

The majority of the new developments reported here aim to enhance the rate of colouration in bipyridilium-based ECDs.

8.4.1 Modulated Light Scattering

Barclay *et al.* [60] have enhanced the rate of colouration by using so-called 'metal-ion catalysis' (the identity of the metal not being stated). An intense optical contrast is achieved using external optics such as a complex array of lenses and an ellipsoidal mirror [60]. Although the effect was not wholly chemical, the extent of intensity change is expressed as a contrast ratio: the *'CR'* quoted for an HV-based device was 60:1, generated in 1 ms. The film of $HV^{+\bullet}$ was electrodeposited with as little as 1 mC cm^{-2} (bleaching of this image took 10 ms).

8.4.2 Pulsed Potentials

Pulses of current have been shown to enhance the rate at which electrochromic colour is formed, relative to colouration with a continuous potential [91]. The procedure relies on reaction of $bipm^{2+}$ (from the bulk solution) with $bipm^{o}$ electrogenerated during the current pulse. The reaction is comproportionation, eq. 8.4, so a fairly cathodic potential must be applied at the working electrode. The amounts of $bipm^{2+}$ and $bipm^{o}$ at the electrode and in the region around the electrode depleted of $bipm^{2+}$ will govern the rate of comproportionation and hence the rate of product colour formation. Thus for a given $[bipm^{2+}] + [bipm^{o}]$ in such a region, the most intense colour will ensue when the two species are in equal concentration. It is envisaged that the pulse procedure engenders this equality, or at least a close approximation to it.

8.4.3 Polyelectrochromism

Bipyridilium salts may typically possess three colours, one for each oxidation state in Fig. 8.1, although the dication in solution is essentially colourless. Viologen electrochromes comprising *n*-bipyridilium units may thus, in principle, exhibit $2n + 1$ colours. This approach has been utilised several times using several bipyridilium units either connected with alkyl linkages [92, 93] or benzylic moieties [94]. Also note, juxtaposition of bipyridilium and Prussian blue electrochromes enables a five-colour ECD to be made [95]. Polyelectrochromic species are collated in chapter 11.

References

[1] H. Kamagawa and T. Suzuki, *J.C.S., Chem. Commun.*, (1985) 525.

[2] N.S. Scaricifti, M. Mehring and N. Neugebauer, *Synth. Met.*, 41–43 (1991) 2971.

[3] L. Michaelis and E.S. Hill, *J. Gen. Physiol.*, 16 (1933) 859.

[4] L. Michaelis, *Chem. Rev.*, 16 (1935) 243.

[5] L.A. Summers, *Adv. Hetero. Chem.*, 35 (1984) 281.

[6] A.J. Bard, A. Ledwith and H.J. Shine, *Adv. Phys. Org. Chem.*, 13 (1976) 155.

[7] C.L. Bird and A.T. Kuhn, *Chem. Soc. Rev.*, 10 (1981) 49.

[8] D.R. Rosseinsky and P.M.S. Monk, *J.C.S., Faraday Trans.*, 89 (1993) 219.

[9] D.R. Rosseinsky and P.M.S. Monk, *J.C.S., Faraday Trans.*, 90 (1994) 1127.

[10] B. Emmert and O. Varenkamp, *Chem. Berichte*, 56 (1923) 490.

[11] G. Lavey and T.W. Emmertson, *J. Phys. Chem.*, 87 (1983) 829.

[12] R.E. van der Leest, *J. Electroanal. Chem.*, 43 (1973) 257.

[13] E.M. Kosower, 'An Introduction to Physical Organic Chemistry', Wiley, New York, 1968.

[14] R.N.F. Thorneley, *Biochim. Biophys. Acta*, 333 (1974) 487.

[15] T. Watanabe and K. Hondo, *J. Phys. Chem.*, 86 (1982) 2617.

[16] B. Emmert, *Chem. Berichte*, 53 (1920) 370.

[17] K. Belinko, *Appl. Phys. Lett.*, 29 (1976) 363.

[18] E. Müller and K.A. Bruhn, *Chem. Berichte*, 86 (1953) 1122.

[19] J.E. Carey, J.E. Cairns and J.E. Colchester, *J.C.S., Chem. Commun.*, (1969) 1290.

[20] E. Kosower and J.L. Cotter, *J. Am. Chem. Soc.*, 86 (1964) 5524.

[21] W. Schartz, Ph.D thesis, University of Wisconsin, (1962). Quoted by Kosower in reference [20] above.

[22] F.M. Hawkridge and J.F. Stargardt, *Analytica Chimica Acta*, 146 (1983) 1.

[23] S. Imabayashi, N. Kitamura, S. Tazuke and K. Tokuda, *J. Electroanal. Chem.*, 243 (1988) 143.

[24] *New Electronics*, 7 (1986) 66.

[25] P.M.S. Monk, N.M. Hodgkinson and S.K. Ramzan: unpublished results.

[26] F. Müller and S.G. Mayhew, *Biochem. Soc. Trans.*, 10 (1982) 176.

[27] P.M.S. Monk, Ph.D thesis, University of Exeter, 1990.

[28] H. Mori and J. Mizuguchi, *Jpn. J. Appl. Phys.*, 26 (1987) 1356.

[29] M.S. Wrighton and D.C. Bookbinder, *J. Electrochem. Soc.*, 130 (1983) 1081.

[30] M.S. Wrighton and C.-F. Shu, *J. Phys. Chem.*, 92 (1988) 5221.

[31] M.S. Wrighton, R.N. Dominey and T.J. Lewis, *J. Phys. Chem.*, 87 (1983) 5345.

[32] K. Itaya, H. Akahoshi and S. Toshima, *J. Phys. Chem.*, 85 (1981) 818.

[33] M.S. Simon and P.T. Moore, *J. Polym. Sci. Polym. Chem. Ed.*, 13 (1975) 1.

[34] J.-C. Moutet, L. Coche and A. Deronzier, *J. Electroanal. Chem.*, 198 (1986) 187.

[35] A.A. Berlin, L.V. Zherebtsova and Y.F. Rabazobovskii, *Polym. Sci. (USSR)*, 6 (1964) 67.

[36] M.A. Stile, Ph.D. thesis, University of Warwick, 1986.

[37] A. Factor and G.E. Heisolm, *Polym. Lett.*, 9 (1971) 289.

[38] H. Sato and T. Tamamura, *Jpn. J. Appl. Phys.*, 24 (1979) 2075.

[39] K.W. Willman and R.W. Murray, *J. Electroanal. Chem.*, 133 (1982) 211.

[40] A.F. Sammells, *Government Reports and Announcements Index US*, 87 (1987) Abstract No. 703,869, cited in *Chem. Abstr*. 107: 86,064m.

[41] A.F. Sammells and N.U. Pujare, *J. Electrochem. Soc.*, 133 (1986) 1270.

[42] J.M. Calvert, T.J. Manuccia and R.J. Nowak, *J. Electrochem. Soc.*, 133 (1986) 951.

[43] C.J. Schoot, J.J. Ponjée, H.T. van Dam, R.A. van Doorn and P.J. Bolwijn, *Appl. Phys. Lett.*, 23 (1973) 64.

[44] Philips Ltd., British Patent 1,302,000, (1971).

[45] J.G. Kenworthy, ICI Ltd., British Patent, 1,314,049, (1973).

[46] H.T. van Dam and J.J Ponjée, *J. Electrochem. Soc.*, 121 (1974) 1555.

[47] R.G. Compton, P.M.S. Monk, D.R. Rosseinsky and A.M. Waller, *J.C.S., Faraday Trans.*, 86 (1990) 2583.

[48] R.J. Jasinski, *J. Electrochem. Soc.*, 124 (1977) 637.

[49] J. Bruinink and P. van Zanten, *J. Electrochem. Soc.*, 124 (1977) 1232.

[50] R.J. Jasinski, *J. Electrochem. Soc.*, 125 (1978) 167.

[51] J. Bruinink and C.G.A. Kregting, *J. Electrochem. Soc.*, 125 (1978) 1619.

[52] T. Berzins and P. Delahay, *J. Am. Chem. Soc.*, 75 (1953) 555.

[53] R.J. Jasinski, *J. Electrochem. Soc.*, 126 (1978) 1619.

[54] R.G. Barradas, S. Fletcher and L. Duff, *J. Electroanal. Chem.*, 100 (1979) 759.

[55] G.G. Barna, *J. Electrochem. Soc.*, 127 (1980) 1317.

[56] N.J. Goddard, A.C. Jackson and M.G. Thomas, *J. Electroanal. Chem.*, 159 (1983) 325.

[57] A. Bewick, A.C. Lowe and C.W. Wederell, *Electrochim. Acta*, 28 (1983) 1899.

[58] A. Bewick, D.W. Cunningham and A.C. Lowe, *Makromol. Chem., Macromol. Symp.*, 8 (1987) 355.

[59] D.R. Rosseinsky, P.M.S. Monk and R.A. Hann, *Electrochim. Acta*, 35 (1990) 1113.

[60] D.J. Barclay, B.F. Bowden, A.C. Lowe and J.C. Wood, *Appl. Phys. Lett.*, 42 (1983) 911.

[61] A.J. Bard and A.E. Kaifer, *J. Phys. Chem.*, 89 (1985) 4876.

[62] D.R. Rosseinsky and P.M.S. Monk, *Solar Energy Mater. Solar Cells*, 25 (1992) 205.

[63] B. Scharifker and C. Wehrmann, *J. Electroanal. Chem.*, 185 (1985) 93.

[64] A. Golden and J. Przyluski, *Electrochim. Acta*, 30 (1985) 1231.

[65] D.R. Rosseinsky and P.M.S. Monk, *J.C.S., Faraday Trans.*, 86 (1990) 3597.

[66] P.M.S. Monk, R.D. Fairweather, J.A. Duffy and M.D. Ingram, *J.C.S., Perkin Trans. II*, (1992) 2039.

[67] O. Poizat, C. Sourisseau and J. Corset, *J. Molec. Struct.*, 143 (1986) 203.

[68] R.G. Compton, P.M.S. Monk, D.R. Rosseinsky and A.M. Waller, *J. Electroanal. Chem.*, 267 (1989) 309.

[69] B. Beden, O. Enea, C. Lamy and F. Hahn, *J. Electroanal. Chem.*, 170 (1984) 357.

[70] P. Crouigneau, O. Enea and B. Beden, *J. Electroanal. Chem.*, 218 (1987) 307.

[71] F.-R.F. Fan, B. Reichman and A.J. Bard, *J. Electrochem. Soc.*, 127 (1980) 333.

[72] P. Crouigneau, O. Enea and C. Lamy, *Nouv. J. Chem.*, 10 (1986) 539.

[73] A. Yasuda, H. Kondo, M. Itabashi and J. Seto, *J. Electroanal. Chem.*, 210 (1986) 265.

[74] T. Lu and T.M. Cotton, *J. Phys. Chem.*, 91 (1987) 5978.

[75] M. Osawa and W. Suetaka, *J. Electroanal. Chem.*, 270 (1989) 261.

[76] T. Sawada and A.J. Bard, *J. Photoacoustics*, 1 (1982/3) 317.

[77] R.E. Malpas and A.J. Bard, *Anal. Chem.*, 52 (1980) 109.

[78] G.H. Brilmyer and A.J. Bard, *Anal. Chem.*, 52 (1980) 685.

[79] G.S. Ostrom and D.A. Buttry, *J. Electroanal. Chem.*, 256 (1988) 411.

[80] D.R. Rosseinsky, J.D. Slocombe, A. Soutar, P.M.S. Monk, and A. Glidle, *J. Electroanal. Chem.*, 259 (1989) 233.

[81] M. Datta, R.E. Jansson and J.J. Freeman, *Appl. Spectrosc.*, 40 (1986) 251.

[82] A. Yasuda, H. Mori, Y. Takehana and A. Ohkoshi, *J. Appl. Electrochem.*, 14 (1984) 323.

[83] A. Nakahara and J.H. Wang, *J. Phys. Chem.*, 67 (1963) 496.

[84] A.S.N. Murthy and A.P. Bhardwaj, *Spectrochimica Acta.*, 38A (1982) 207.

[85] S.F. Kranarenko and I.P. Krainov, *Ukr. Khim. Zh.*, 51 (1985) 501, (in Russian), cited in *Chem. Abstr.* 103 (1985) 112,232.)

[86] J. Bruinink, J.J Ponjée and C.G.A. Kregting, *J. Electrochem. Soc.*, 124 (1977) 1854.

[87] G.G. Barna and J.G. Fish, *J. Electrochem. Soc.*, 128 (1981) 1290.

[88] J.A. Barltrop and A.C. Jackson, *J.C.S., Perkin Trans. II*, (1984) 367.

[89] D.J. Barclay, C.L. Bird, D.H. Kirkman, D.H. Martin and F.T. Moth, *Society for Information Displays International Conference Digest*, XI (1980) 124.

[90] D.J. Barclay and D.H. Martin, in E.R. Howells (ed.), 'Technology of Chemicals and Materials for the Electronics Industry', Ellis Horwood, Chichester, 1984. Chapter 15.

[91] P.M.S. Monk, R.D. Fairweather, M.D. Ingram and J.A. Duffy, *J. Electroanal. Chem.*, 359 (1993) 301.

[92] Z. Porat, Y.M. Tricot, I. Rubinstein and B. Zinger, *J. Electroanal. Chem.*, 315 (1991) 217.

[93] Z. Porat, Y.M. Tricot, I. Rubinstein and B. Zinger, *J. Electroanal. Chem.*, 315 (1991) 225.

[94] D.R. Rosseinsky and P.M.S. Monk, *J. Appl. Electrochem*, 24 (1994) 1213.

[95] R.J. Mortimer, *J. Electrochem. Soc.*, 138 (1991) 633.

9 Electroactive Conducting Polymers

9.1 Introduction

Many aromatic species form solid conducting polymers following electron transfer. Exemplar monomers include aniline, pyrrole or thiophene which form polyaniline, polypyrrole and polythiophene respectively. The heterocycle may be polymerised chemically, for example using a radical initiator like dibenzoyl peroxide; alternatively, polymers may be prepared electrochemically. Pyrrole monomer in MeCN solution (with electrolyte) polymerises to form a film of polypyrrole on an electrode at positive [1] potentials. Such films comprise poly-carbocations 'doped' with counter anions. All conducting and redox-active polymers are potentially electrochromic in thin-film form.

The name 'doping' has been criticised by Wegner [2] as an inaccurate term for the movement of counter-ions and charge, but the name is now so widely used that new nomenclature would only confuse.

Doped polymers typically possess a high conductivity whereas the corresponding undoped (neutral) redox forms are insulating. Such conductivity follows from the extended conjugation within the polymer, longer chains promoting high conductivity in doped polymers. The average number of linked monomer units within a conducting polymer is often called a 'conjugation length' in the literature.

Conducting polymers are type 3 electrochromes since they are permanently solid. Redox changes in the polymer introduce doping with counter ions and, since new energy levels are filled or become vacant, new optical absorption bands are formed in accompaniment with transfer of the electron. Polypyrrole-type films exhibit electrochromism because the oxidised (doped) and reduced (undoped) states show different colours.

In this work, polymers will be named as poly(monomer) except for the well-known examples of polyaniline, polypyrrole and polythiophene.

The field of conducting-polymer electrochromes has been reviewed before. Recent works include 'Application of Polyheterocycles to Electrochromic Display Devices' by Gazard [3] (1986), 'Electrochromic Devices' by Mastragostino [4] (1993) and 'Electrochromism of Conducting Polymers' [5] by Hyodo (1994). Other less substantial, but informative works include 'Chromogenic Materials (Electrochromic)' by Greenberg [6], while 'Conducting Polymers and Plastics' by Margolis [7] and an article by Mortimer [8] have short sections on electrochromic displays.

9.2 Polyaniline Electrochromes

Anodic polymerisation of colourless aniline in solution yields polyaniline. The best films are prepared galvanostatically from aqueous acid, for example, Yoneyama *et al.* [9] used 1 M aniline in 2 M HCl solutions and a current density of 0.1 mA cm^{-2}.

The compositions of the various redox states are indicated in Fig 9.1 which is based on references [10] and [11].

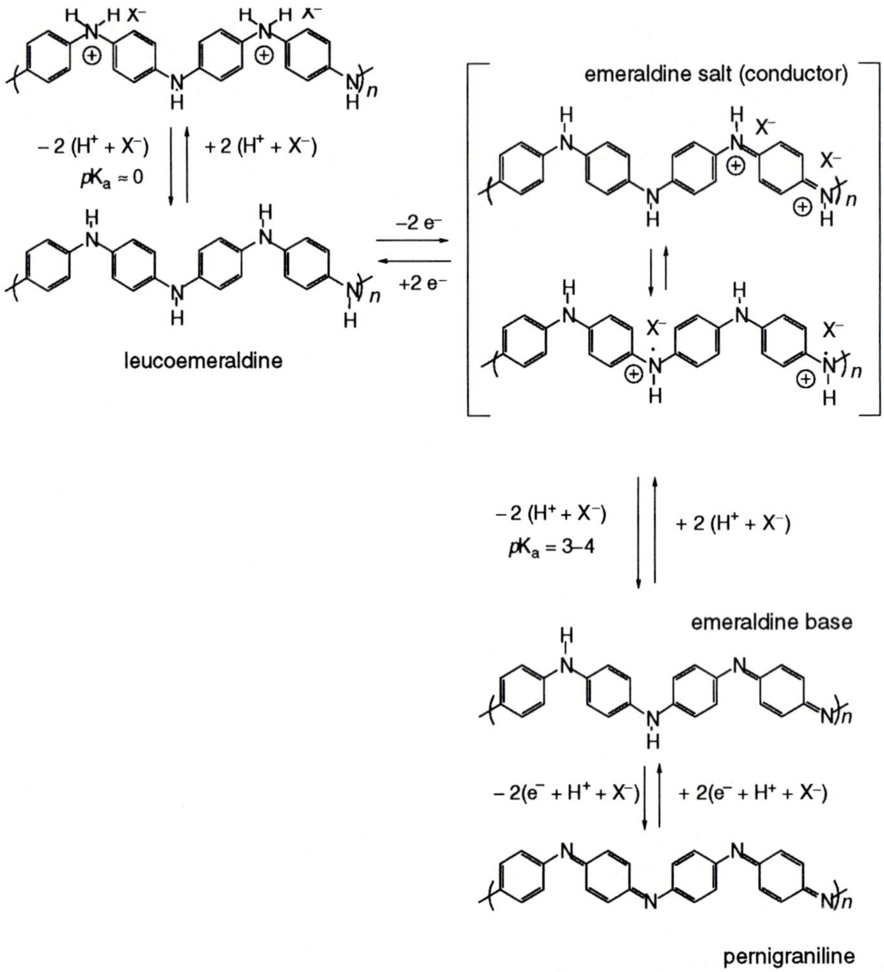

Fig. 9.1 The structures and names of the various polyaniline redox states.

Leucoemeraldine is an insulator since all rings are benzenoid in form and separated by –NH– or $-NH_2^+-$ groups, thus preventing conjugation between rings. Emeraldine as either base or salt has a ratio of three benzenoid rings to one quinoidal ring. This form of polyaniline is electrically conductive. Pernigraniline has equal proportions of quinoidal and benzenoid moieties and shows metallic conductivity. The aniline units within the polyaniline backbone are not coplanar, as has been shown by solid-state ^{13}C-NMR [12].

Electrodes bearing such polyaniline films are polyelectrochromic and exhibit the following reversible colour changes as the potential is varied (table 9.1): transparent leucoemeraldine to yellow/green 'emeraldine' to dark blue/black 'pernigraniline', in the potential range −0.2 V to +1.0 V [9].

Pernigraniline is an intense blue colour but appears black at very positive potentials if the film is thick. The yellow form of polyaniline has an absorbance maximum at 305 nm, but no appreciable absorbance in the visible region. Spectroscopic data for the redox forms of polyaniline are summarised in table 9.1. Fig. 9.2 shows the absorbance spectrum as a function of potential. It can be seen that the wavelength maximum is a function of oxidation state, λ decreasing as the extent of oxidation increases.

Kobayashi *et al.* [13] state that another form of polyaniline exists between yellow and green, at potentials less positive than +0.3 V.

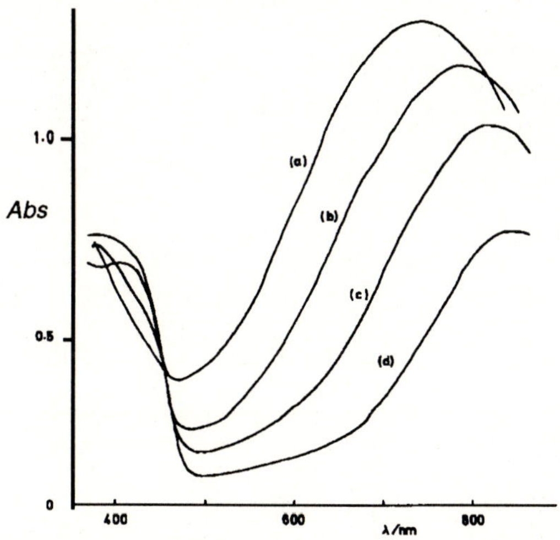

Fig. 9.2 *UV-vis spectrum of polyaniline as a thin film on ITO (vs ITO as a blank) in 0.1 M HCl solution. The potential of the underlying electrode was (a) 2.0 V (b) 1.0 V (c) 0.9 V and (d) 0.7 V (unpublished data).*

The mechanism of electropolymerisation proceeds via a radical cation [14, 15]; scheme 9.1 (which is simplified from ref. [15]) depicts the chain initiation process. Note that aniline may in fact be protonated since strongly acidic solutions are used, and polyaniline is not formed in such good yield at higher pH.

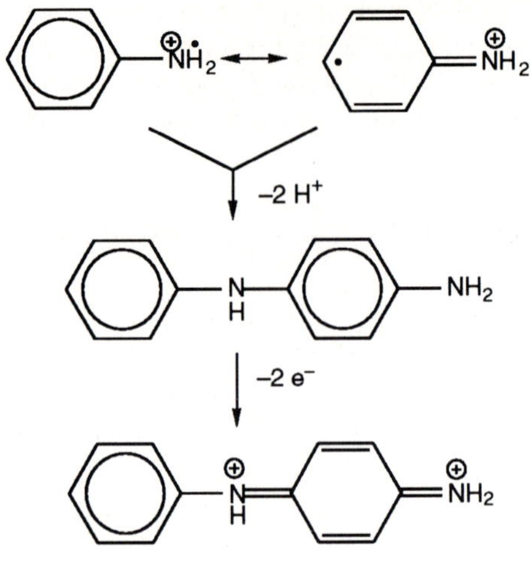

Scheme 9.1

Here the oxidised (blue) polyaniline is the initially-formed product. The stablest polyaniline ECDs operate over a restricted potential range to allow only the yellow and green states. Such a system has a cycle life in excess of 10^6 cycles [9]. The technology of the electrochromism of polyaniline in ECDs has been patented [17].

Table 9.1 Colours, Wavelength Maxima and Potential Range in Which Polyaniline Redox Species are Observed.

Colour	λ_{max}/nm	Potential Range/V	Comments	Ref.
yellow	305	−ve of −0.2	Undoped	[9]
green	740	+ve of 0.0		[9]
blue	420	+ve of +0.6		[16]
black	$740 < \lambda < 840$	+ve of +1.0	Fully doped	[16]

Other studies of polyaniline have been performed with photo-thermal spectroscopy [18], ellipsometry [19] and reflectance Raman spectroscopy [20–25], for example, by Hugot-Le Goff and co-workers. It is concluded that films are not stable in strongly acidic milieu, chain cleavage occurring in tandem with crosslinking between chains [25]. Although the best polyaniline films are *prepared* in solutions of concentrated acid, electrochromic operation should for this reason be performed at a pH higher than about 2 or 3.

The ellipsometry data [19] reveal that polyaniline films expand by up to 30% on redox cycling. There is also considerable hysteresis in film thickness.

The electrochromic properties of chemically (rather than electrochemically) formed polyaniline has been described by Mastragostino and co-workers [26]. Chemical oxidants used in their study included KIO_3, $(NH_4)_2S_2O_8$ and $K_2Cr_2O_7$. An alternative method of oxidatively forming polyaniline is to warm aniline monomer with $FeCl_3$ [27].

9.2.1 Polymers Derived from Substituted Anilines

While the electrochromism of polyaniline has received much attention [9–26, 28] *N*-substituted anilines have not been studied so often, although the first polymer prepared from a substituted aniline was poly(*o*-toluidine), made chemically by Green and Woodhead in 1910 [29]. Electrochemically prepared poly(*o*-toluidine) has a shorter polymer chain than does polyaniline [30].

Several polymers of substituted anilines have been prepared in MacDiarmid's laboratory [30], for which spectral properties are listed in table 9.2.

Table 9.2 Wavelength Maxima of the Base Forms of Poly(Substituted Aniline) in DMF Solution [30].

Aniline Monomer	λ_{max}/nm
aniline	330 and 620
o-toluidine	318 and 600
o-ethylaniline	311 and 600

When an aniline has an anionic moiety at nitrogen, the resultant polymer is often said to be 'self doped' since no additional (free) anions need be incorporated within the polymer during redox cycling. Electrochemically induced polymerisation of 4-anilino-1-butane-sulphonic acid (as a sodium salt) results in the formation of the polymer (**I**), which is water soluble thus allowing films to be cast [31]. The cycle life is of polymer (**I**) reported to be higher than for polyaniline alone [31].

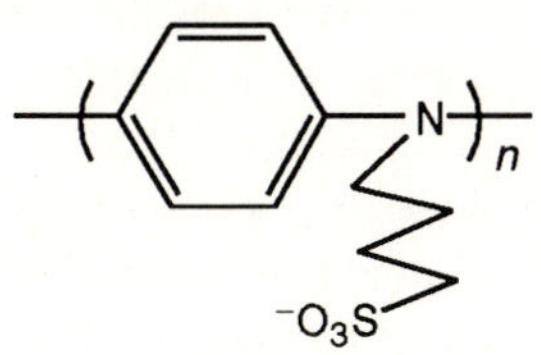

I

2-Chloroaniline may be electropolymerised from non-aqueous dichloroethane solutions containing tetrabutylammonium tetrafluoroborate rather than a protonic acid [32] (the perchlorate salt is rather insoluble). The resultant polymer is, apparently, soluble.

Poly(*o*-phenylenediamine) is electrochromic [33], as are the polymers formed by electrochemical coupling of of 3-chlorophenylenediamine [34] and 5-amino-1-naphthol [35, 36]. Polymers derived from diphenylamines include poly(*N*-alkyldiphenylamine) [37], poly(3-methyldiphenylamine) and poly(3-methoxy-diphenylamine) [38]. Other poly(di-arylamines) have been prepared by Guay *et al.* [39].

Aminoquinones, such as 5-aminonapthoquinone [40], form electrochromic polymers at very positive potentials, as detailed in section 10.1.3.

9.2.2 Polymers Derived from Other Aromatic Amines

1-pyreneamine (**II**) may be electropolymerised [41]: the polymeric electrochrome was generated by taking an electrode pre-coated with monomer and applying a positive potential following immersion in an electrolyte solution. The monomer film was prepared by vacuum evaporation onto the electrode.

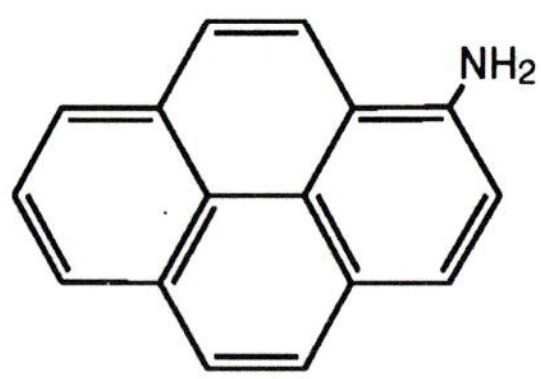

II

Naphthidine (1,1'-binaphthalene-4,4'-diamine) may also be polymerised anodically to form conducting, electrochromic thin films [42].

9.2.3 Composite Polyaniline Materials

Polymers obtained by electropolymerising aniline in the presence of other (possibly) electrochromic materials have often been prepared, Table 9.3. The electrochromic properties are reportedly superior to those of polyaniline alone. Exemplifying the marked change in properties possible, a composite of polyaniline and poly(styrene sulphonic acid) is soluble in water [43]; the only solvent able to dissolve polyaniline itself is concentrated sulphuric acid. Much of the work in table 9.3 derives from the laboratory of Morita.

Table 9.3 Examples of Composite Electrochromes Based on Polyaniline or Poly(*o*-phenylenediamine).

Additive	Ref.
cellulose acetate	[44]
methylene blue	[45]
nitrilic rubber	[46]
polyion complexes	[47]
polythiophene	[48]
poly(3-methylthiophene)	[48]
poly(styrene tetrasulphonate)	[43, 49]
poly(vinyl alcohol)	[50, 51]
TiO_2	[52]
WO_3	[53]
[a]Prussian blue	[33]

[a]In poly(*o*-phenylenediamine); all others in polyaniline.

9.3 Polypyrrole Electrochromes

Polypyrrole was first prepared electrochemically in 1965 by MacNeill and co-workers [54]. The mechanism of electropolymerisation involves a radical-cation intermediate [1], scheme 9.2 [55].

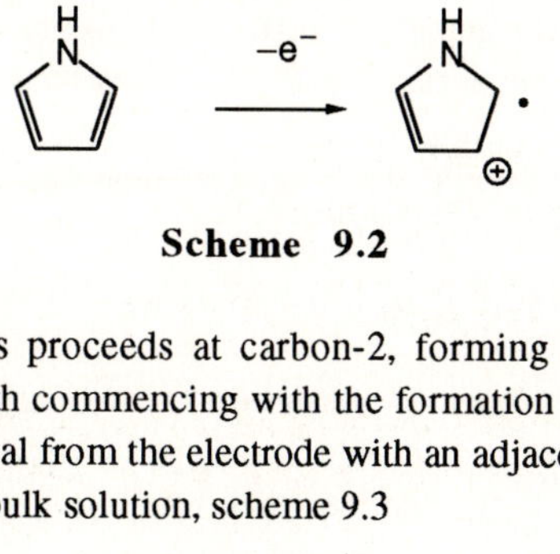

Scheme 9.2

Coupling of such radicals proceeds at carbon-2, forming polypyrrole or oligomer, presumably polymer growth commencing with the formation of 2,2-dipyrrole following electrophilic attack of radical from the electrode with an adjacent pyrrole species (neutral or radical states) from the bulk solution, scheme 9.3

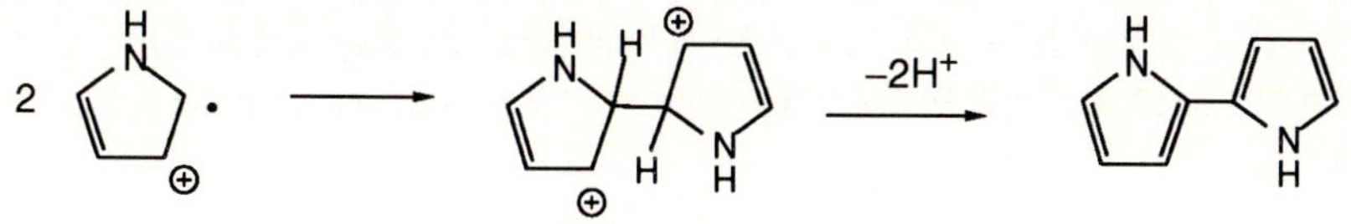

Scheme 9.3

Polymer growth is known to proceed following nucleation at the electrode–solution interface [56], that is, electropolymerisation is a heterogeneous reaction. The polymer is formed in its oxidised form [57].

X-ray diffraction of pyrrole oligomers suggests the polymer to be co-planar [58] but that substitution at nitrogen, and at carbon-3, causes the polymer to have significant twist in its backbone (the dihedral angle ϕ is non-zero), Fig. 9.3.

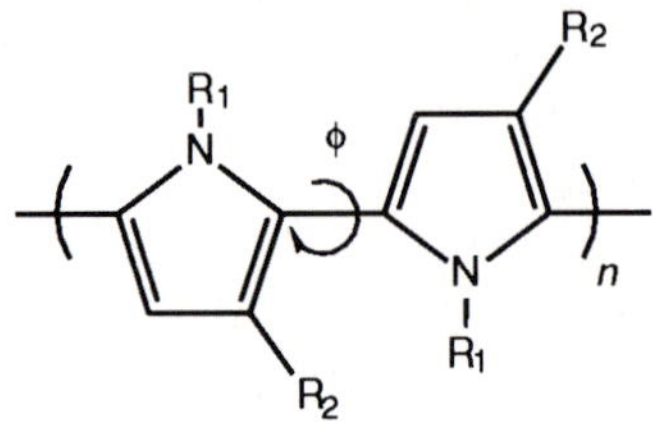

$\phi \neq 0$ if:

$R_1 \neq H$ and
$R_2 \neq H$

Fig. 9.3 Segment of polypyrrole chain showing the dihedral angle ϕ.

The first reports of the electrochemical preparation of thin-film polypyrrole were those of Diaz *et al.* in 1979 [1, 59, 60]. The nature of the polymer product obtained by electro-oxidising pyrrole monomer depends greatly on the conditions used during electropolymerisation, for example, the potential used for deposition [61], the temperature [62], the presence of specific acids and bases [55], the solvent [61, 63] and the counter ion present in the pyrrole-containing solution [64–66]. The effect of counter ion is illustrated by some of the polymers in table 9.4 (a).

Removal of all dopant anions from polypyrrole yields a pale yellow film. However, *complete* de-doping is only achieved if films are extremely thin. This means that polypyrrole of thickness commensurate with ECD construction – say, thicker than about 1 µm – is strongly coloured in both its doped and undoped forms, and hence has a low *CR*. The usual 'undoped' polypyrrole is yellow/green ($\lambda_{max} = 420$ nm) while doped (oxidised) polypyrrole is blue/violet ($\lambda_{max} = 670$ nm) [67], although Diaz [68] quotes $\lambda_{max} = 406$ nm for the reduced polymer and $\lambda_{max} = 530$ nm for the doped (oxidised) form.

Polypyrrole doped with dodecylbenzenesulphonate anion [69] and dodecylsulphate anion [70, 71] have been prepared, the later having a cited cycle life of 10^4 cycles.

The kinetics of polymer degradation have been studied by Park *et al.* [72]. The degradation rate is proportional to the concentration of the acid used during redox cycling and also depends on the potential used for deposition, films formed at more anodic potentials being more susceptible to degradation.

9.3.1 Polymers Derived from Substituted Pyrroles

A product with improved electrochromic properties is that formed by polymerising 3,4-disubstituted pyrroles. For example, the polymer product of 3-methyl-4-carboxy-pyrrole [73] has an enhanced cycle life and *CR* (λ_{max} (reduced state) = 406 nm and λ_{max} (oxidised state) = 530 nm). Poly(*N*-methylpyrrole) is also said to have improved electrochromic characteristics [74] with respect to the polymer of unsubstituted pyrrole, and has λ_{max} = 580 nm for the absorbing (oxidised) form of the polymer. Spectra of poly(*N*-methylpyrrole) as a function of potential are given by Yaniger and Vidrine [75]. Poly(*N*-trimethylsilylpyrrole) has been reported [76].

Table 9.4 (b) gives electrochemical and spectroscopic data for substituted pyrroles.

Table 9.4 Properties of Pyrrole-based Polymers Formed Electrochemically from MeCN solution (after ref. [3]).

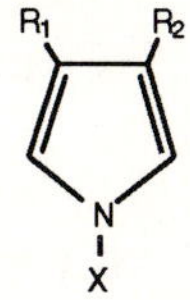

(a) Electrochemical Properties from CVs Obtained at a Scan Rate of 100 mV s^{-1}.

X	R_1	R_2	Electrolyte in MeCN	E_{pa}/V	E_{pc}/V	Ref.
H	H	H	$LiClO_4$	−0.08	−0.40	[77]
			TEAP	−0.18	−0.38	[77]
			TBAT	−0.12	−0.29	[77, 78]
			TBAT	−0.15	−0.25	[79]
H	H	$COCH_3$	TBAT	0.90	0.75	[3]
H	CH_3	CH_3	TBAT	0.10	−0.10	[3]
CH_3	H	H	TBAT	0.46	0.39	[78]
CH_3	H	H	TBAT	0.40	0.50	[79]
Ph	H	H	TBAT	0.60	0.74	[79]

(b) Electrochromic Properties (TBAT in MeCN).

X	R_1	R_2	Polymer Colour Oxidised Form	Polymer Colour Reduced Form	Potential Range/V	η/cm^2 C^{-1} (λ_{max}/nm)
H	H	H	brown	yellow	0 to 0.7	100 (500)
H	H	$COCH_3$	yellow/brown	brown/yellow	0 to 1.1	100 (580)
H	CH_3	CH_3	purple	green	−0.5 to 0.5	20 (440)
CH_3	H	H	brown/red	orange/yellow	0 to 0.8	20 (500)

9.3.2 Polymers Derived from Pyrrole Analogues

Electrochromic polymers are obtained by anodic electron transfer to indole (**III**) [80, 81] or *N*-methyl*iso*indole [82].

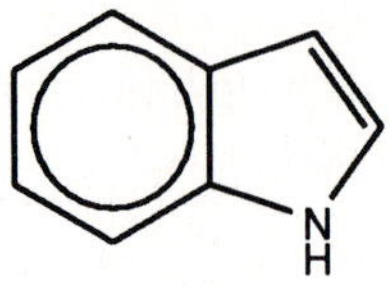

III

9.3.3 Composite Polypyrrole Electrochromes

If a solution of pyrrole (or pyrrole-based heterocycle) contains an additive during anodic polymerisation, the resultant polymer composite incorporating the additive has altered electrochromic properties from those of the polymer alone; notably, the electrical conductivity and electroactivity of films are enhanced [83]. Table 9.5 contains examples of polypyrrole-based composites.

Table 9.5 Examples of Composite Electrochromes Based on Polypyrrole or Poly(dithienopyrrole).

Additive	Ref.
Copper (tetrasulphonated phthalocyanine)	[84]
2,5-dimercapto-1,3,4-thiadiazole	[85]
Fe(TSpc)	[86]
indigo carmine (**IV**)	[87]
indigo carmine	[83]
methylene blue	[88]
$[P_2W_{18}O_{62}]^{6-}$	[89]
$[P_2W_{18}O_{62}]^{6-}$	[90]
$[SiW_{18}O_{62}]^{4-}$	[90]
Tiron catalyst	[91]
WO_3	[92]
poly(vinylpyrrolidone)	[91]
[a]poly(vinyl chloride)	[93]

[a] In poly(dithienopyrrole); all others in polypyrrole.

In many of these studies, the material encapsulated within the polypyrrole matrix is itself electrochromic, thus the dye indigo carmine (**IV**) may act as a redox indicator [83], scheme 9.4. Care is necessary when using such composites, however, because over-oxidation of the polypyrrole causes (**IV**) to be expelled from the film [83].

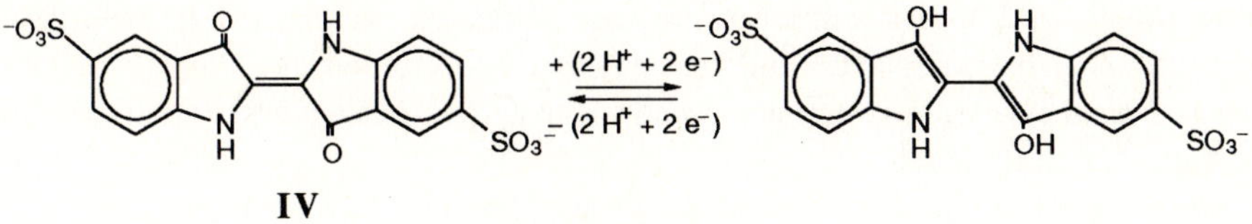

Scheme 9.4

9.4 Polythiophene Electrochromes

9.4.1 Polymers Derived from Thiophene

Polythiophene may be prepared [94] by the oxidative coupling of thiophene at an electrode. Polymerisation proceeds via an anodically generated radical cation [4]:

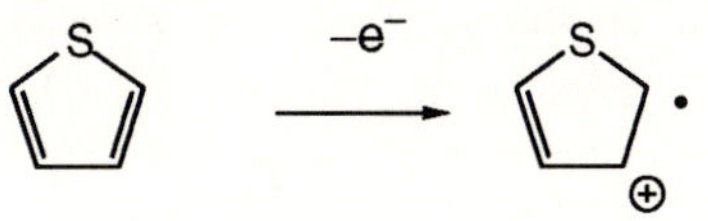

Scheme 9.5

The radical cation can either react with other thiophene units (as charged radical-cation species as in scheme 9.6 or uncharged thiophene):

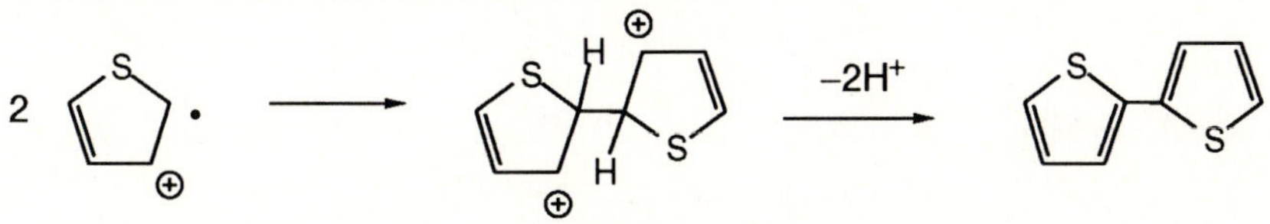

Scheme 9.6

or it can react with incipient polymer, that is, in a chain propagation reaction:

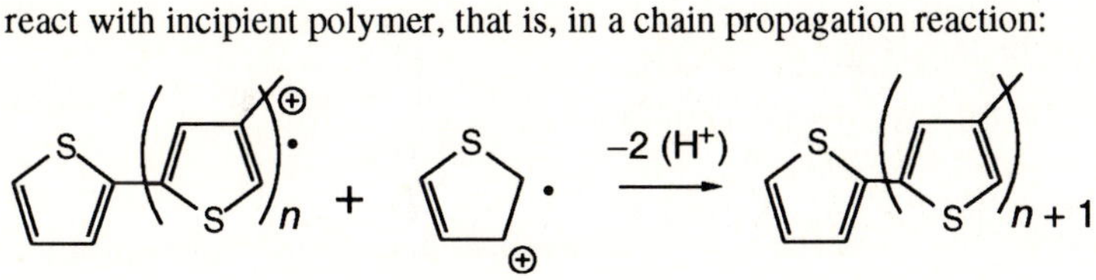

Scheme 9.7

When the oligomer reaches a certain length, the solubility threshold is reached, and solid forms on the electrode. Such nucleation is an important part of the mechanism whereby electrodes are derivatised with thin films of polymer [4].

Films of polythiophene have a relatively low cycle life, evinced after excessive write–erase cycles as a decreased *CR* and an increased resistance to charge transfer. Wang [95], using differential UV-visible spectroscopy, has postulated a mechanism to explain the deactivation of polythiophene films: nucleophilic attack at the ring causes structural changes, chain cleavage and eventual decomposition. The nucleophiles are derived from the solvent or electrolyte.

Finally, many devices have been made which incorporate polythiophene, or one of its derivatives, as the electrochrome. Yoshino used polythiophene doped with diaryliodonium salts for an optical recording and memory device [96]. Optical switching elements [97] and ECDs [98] have also been constructed from polythiophenes.

9.4.2 Polymers Derived from Substituted Thiophenes

The electrochromic properties of polythiophene and of the polymers of several substituted thiophenes are reproduced in tables 9.6 and 9.7, which are based on refs. [67] and [3] respectively. Reference [99] also cites data for poly(3-substituted thiophene)s.

Table 9.6 Polythiophenes: The Effect of Anion on Wavelength Maxima and Oxidation Potential (based on ref. [67]).

Monomer	Anion	λ_{max}/nm	
		Oxidised Form	Reduced Form
thiophene	BF_4^-	730	470
	ClO_4^-	blue	red
	$CF_3SO_3^-$		
3-methylthiophene	BF_4^-	750	480
	ClO_4^-	deep blue	red
	$CF_3SO_3^-$	750	480
	Picryl	blue/green	red
3,4-dimethylthiophene	ClO_4^-	750	620
	$CF_3SO_3^-$	dark-blue	pale brown
2,2'-bithiophene	$CF_3SO_3^-$	680	460
		blue/grey	red/orange

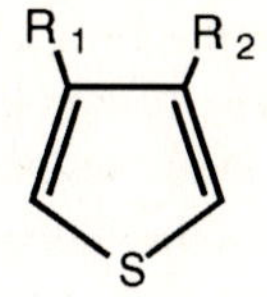

Table 9.7 Properties of Thiophene-based Polymers Formed Electrochemically from MeCN Solution (after ref. [3]).

(a) Electrochemical Properties at a Scan Rate of 100 mV s^{-1}.

R_1	R_2	Electrolyte in MeCN	E_{pa}/V	E_{pc}/V	Ref.
H	H	TBAT	1.10	0.75	[78]
H	CH_3	TBAT	0.90	0.40	[78, 67]
H	CH_3	TBAT	1.10	0.77	[3]
H	Ph	TBAT	1.10	1.00	[3]
Ph	Ph	TBAT	1.25	0.98	[3]
2,2'-bithiophene		$LiClO_4$	1.00	0.82	[77]
		TBAP	1.15	0.60	[77]
		TBAT	1.02	0.60	[77, 78]

(b) Electrochromic Properties (TEAT/MeCN).

R_1	R_2	Polymer Colour Oxidised Form	Polymer Colour Reduced Form	Potential Range/V
H	CH_3	blue	red	0 to 1.1
H	Ph	green/blue	yellow	0.5 to 1.5
Ph	Ph	blue/grey	yellow	0.5 to 0.15
2,2'-bithiophene		blue/grey	red	0 to 1.3

Variation in the length and nature of the substituent at carbon-3 has been investigated often [100–110], for example with alkyl substituents [101, 102] like octyl [100], isopentyl [103] or nonyl [103], or with acetic acid [104], or oxyalkyl (aromatic ether) moeities such as oxyheptyl [103, 105–107] or alkoxy groups [108]. The incorporation of ethers is said to enhance solid-state ionic conductivity [4] since cation coordination is easy, although note that transport numbers of cations such as Li^+ through polyethylene oxide are quite small [109]. Roncali *et al.* [101] have also investigated the polymers formed from 3-substituted thiophenes having both linear or branched alkyl chains. Such polymers, when doped, have absorbance maxima which depend on the nature of the alkyl substituent. Of the thiophene polymers studied, poly(3-methylthiophene) had the best conductivity and durability; for example, the polymer retains over 80% of its original activity after 1.2×10^5 cycles in MeCN [101]. Corradini [110] similarly found that this thiophene produced a superior polymer for ECD inclusion.

For the case of linear alkyl chains at carbon-3, the chain length has an important effect on the electrochromic and polymer-forming properties, heptyl, octyl and nonyl chains imparting the longest conjugation lengths [4, 103]. Table 9.8 illustrates the way optical and electrochemical properties depend on the length of the length of the alkyl chain at carbon-3.

Polymers derived from 3-methoxythiophene are of low molecular weight (i.e. have a short conjugation length) whereas polymers derived from 4,4'-dimethoxy-2,2-bithiophene have longer conjugation lengths [111]. The latter polymer has an electrochromic transition from intense, deep blue when undoped (λ_{max} = 610 nm) to pale grey/blue when oxidatively doped [4, 111].

Gottesfeld *et al.* [112] have prepared a series of polymers of the type (**V**) where R = H, F, Cl, CH_3 or CF_3. Poly(3-trimethylsilylthiophene) [113] is also electrochromic.

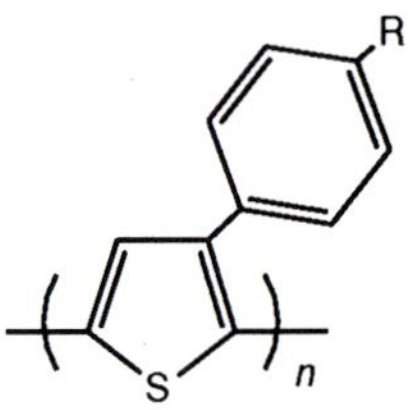

V

Table 9.8 Effect of Chain Length on Optical and Electrochemical Properties of Polymers Derived from 3-Alkylsubstituted Thiophenes (based on ref. [102]).

Substituent at Carbon-3	E^o/V[a]	Polymer Colour by Transmission		λ_{max}/nm[b]
		Reduced Form[c]	Oxidised Form[d]	
$-(CH_2)_{12}CH_3$	0.73	red	blue	481
$-(CH_2)_{13}CH_3$	0.82	red	blue	479
$-(CH_2)_{14}CH_3$	0.86	red	dark blue	485
$-(CH_2)_{15}CH_3$	0.91	brick-red	blue	467
$-(CH_2)_{16}CH_3$	0.94	red	green/blue	455
$-(CH_2)_{18}CH_3$	0.96	dark red	blue	450

Key: [a]All potentials cited *vs* AgCl/Ag in MeCN. [b]λ_{max} refer to reduced (undoped) redox states of films deposited on ITO. [c]Electrode held at −0.2 V. [d]Electrode held at 1.4 V.

9.4.3 Polymers Derived from Oligothiophenes

There has been much interest in the polymers derived from electrochemical oxidation of thiophene-based monomers that comprise more than one thiophene heterocyclic unit. The chemical containing two thiophene units (joined at carbon-2) is called bithiophene while compounds containing three or more thiophene units have the general name of 'oligo-thiophene'.

Mark *et al.* [114] have shown that the wavelength maxima of undoped poly(oligo-thiophene) films decrease as the length of the oligothiophene monomer increases, table 9.9. The oxidation potentials included in this table do not vary much with oligo-thiophene.

Table 9.9 Wavelength Maxima and Oxidation Potentials of Polymers Derived from Oligothiophenes (based on ref. [114]).

Monomer[a]	λ/nm[b]	E_{ox}/V
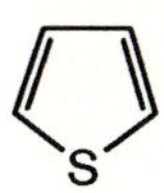	519	0.95
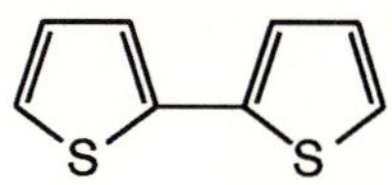	484	1.00
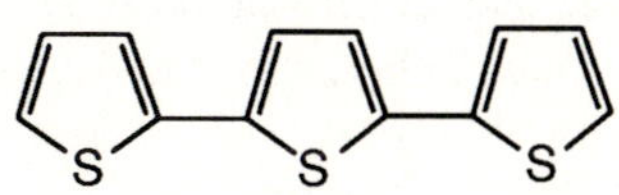	356	1.04
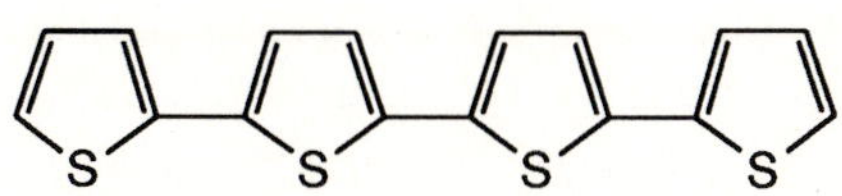	340	0.93

Key: [a]Note that these structures do not represent the molecular stereochemistry. [b]Wavelength maximum refers to the reduced (undoped) redox state of the polymer.

Mastragostino and co-workers [4, 115–118] have prepared a range of polymers based on a core of 3-methylthiophene, but in which the 'monomer' electropolymerised was in fact oligomeric, either comprising linear chains of two or four thiophene units (table 9.10). Polymers were grown by cycling the potential using, for example, ITO electrodes immersed in solutions containing monomer in acetonitrile (and with $LiClO_4$ as electrolyte) [4]. The electrochromic properties of the resultant polymer was strongly dependent on the relative positions of methyl groups on the polymer backbone [4], as shown in Table 9.10.

Table 9.10 Colours of Polymers Derived from Oligomers Based on 3-Methylthiophene (ref. [4]).

Monomer	λ_{max}/nm (undoped)	Polymer Colour Reduced Form	Oxidised Form
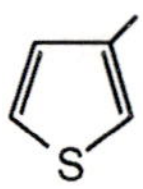	530	purple	pale blue
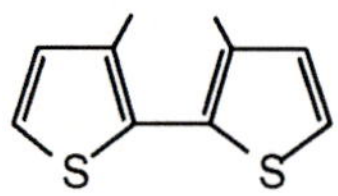	415	yellow	violet
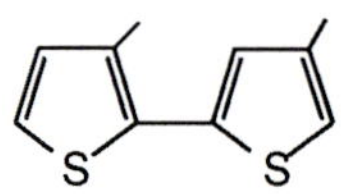	505	red	blue
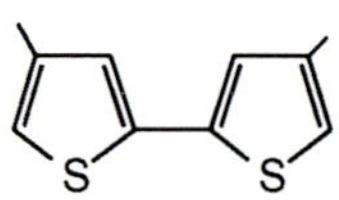	450	orange	blue
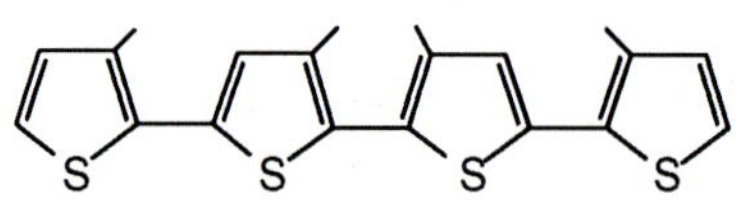	425	yellow	blue
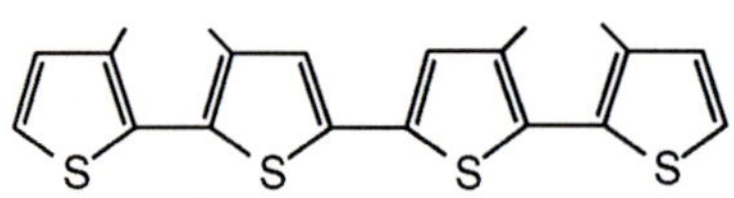	405	yellow	violet
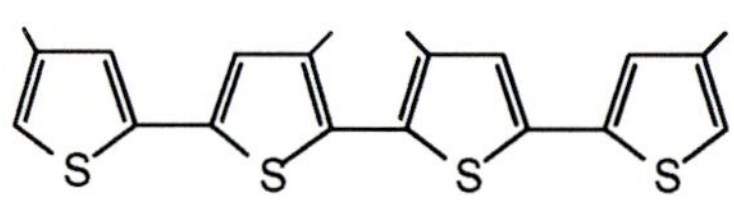	410	yellow	blue/violet
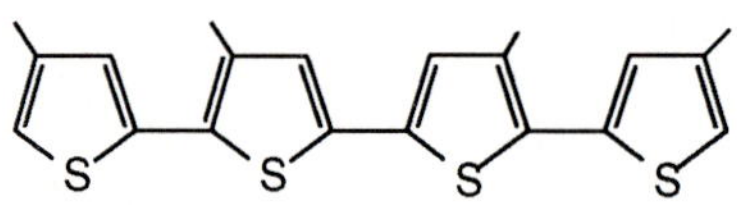	425	yellow/orange	blue

As an adjunct, Mark *et al.* [114] have made oligothiophenes containing alkyl groups at carbon-3 to investigate the effect of dihedral angle ϕ between thiophene planes: groups at carbon-3 cause steric hindrance and bridged species are planar. The results in table 9.11

show that those polymers with the smallest dihedral angles ϕ generally have the highest wavelength maxima. Oxidation potentials are generally unaffected by variations in ϕ.

Table 9.11 Effect of the Dihedral Angle ϕ: Spectroscopic and Electrochemical Characteristics of Poly(oligothiophene)s (based on ref. [114]).

Monomer	λ_{max}/nm (undoped)	E_{ox}/V
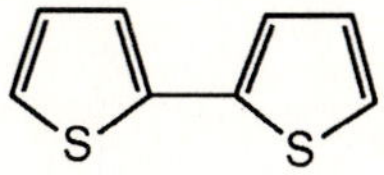	484	1.00
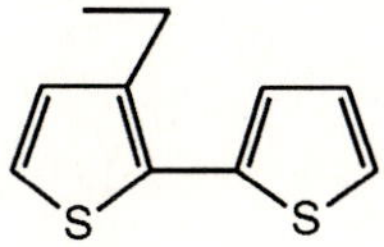	475	0.96
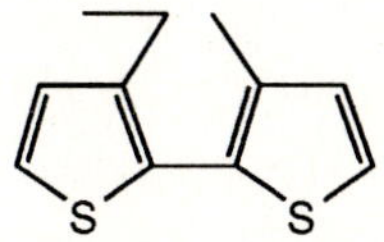	420	0.99
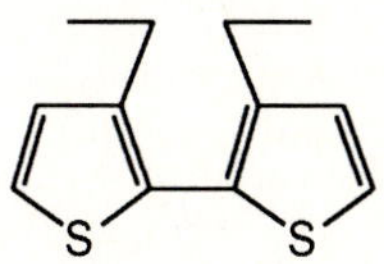	413	0.88
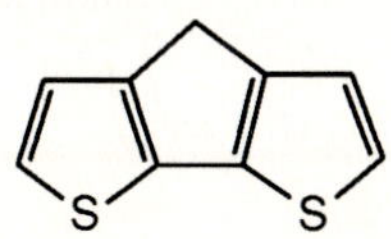	550	0.90
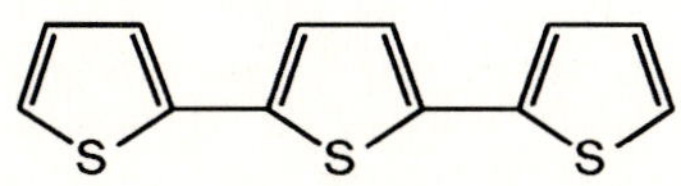	356	1.04
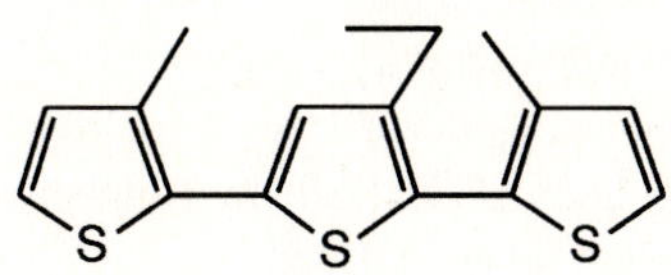	375	0.94

Finally, table 9.12 demonstrates the effect of varying the heteroatom in the chain. While heteroatoms do not cause systematic changes in wavelength maxima, the trends in electronegativity are evident from the relative oxidation potentials [114]. Poly(selenylthiophene) has also been studied by Peulon *et al.* [119].

Table 9.12 The Effect of Varying the Heteroatom within a Polymer Derived from 2-Thieno-(2'-heterocycle) (table based on ref. [114]).

Polymer Repeat Unit	λ_{max}/nm (undoped)	E_{ox}/V
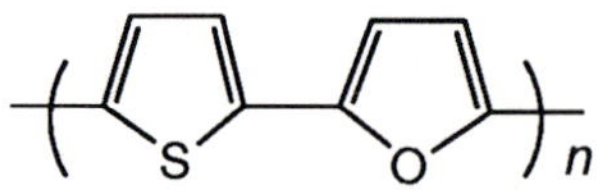	448	0.80
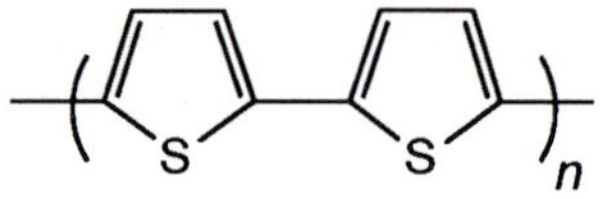	484	1.00
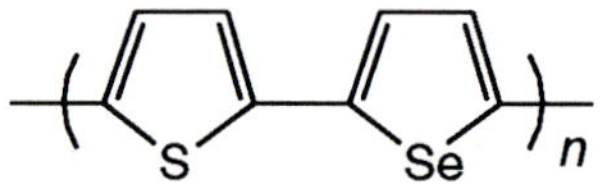	430	1.15

9.4.4 Polymers Derived from *bis*(2-Thienyl) Species

A different thiophene-based monomer has two thiophenes connected via an aromatic (often heterocyclic) link X (**VI**):

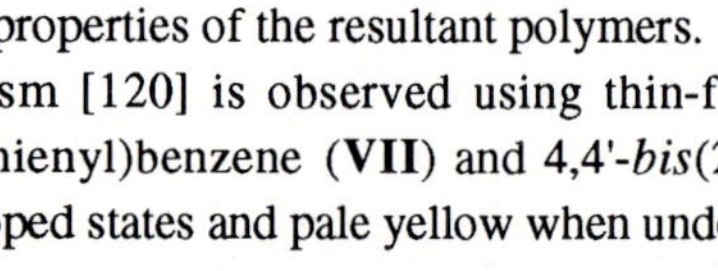

VI

Such species are termed *bis*(2-thienyl)s, and have been prepared by several workers, who report the electrochromic properties of the resultant polymers.

Green electrochromism [120] is observed using thin-films of the two polymers derived from 1,3-*bis*(2-thienyl)benzene (**VII**) and 4,4'-*bis*(2-thienyl)biphenyl (**VIII**), which are green in their doped states and pale yellow when undoped.

VII

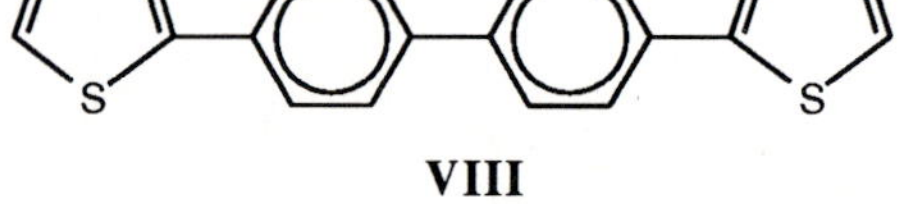

VIII

Similarly, Gottesfeld *et al.* [112] have prepared a range of *bis*(2-thienyl) compounds based on a core of (**IX**) and (**X**). Reference [112] reproduces many spectra but does not list wavelength maxima.

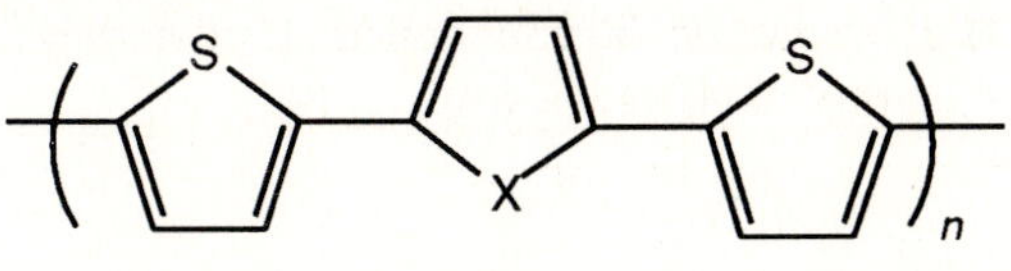

X = S
X = O
X = NH
X = NCH_3
X = $NCH(CH_3)_2$

IX

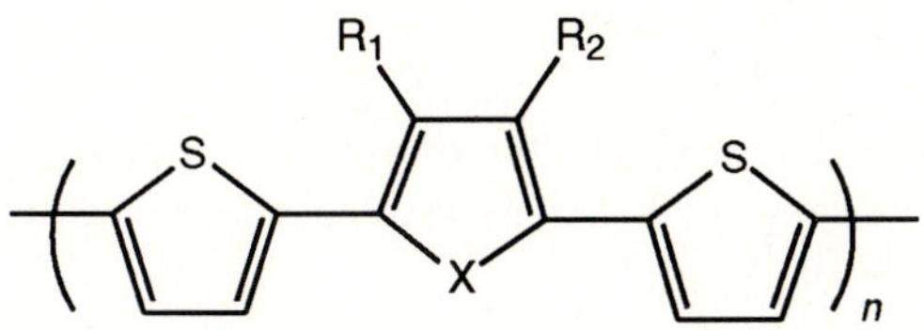

X = NH; $R_{1,2}$ = H
X = NH; $R_{1,2}$ N = CH_3
X = NCH_3; $R_{1,2}$ = H
X = S; R_1 = H; R_2 = CH_3
X = S; $R_{1,2}$ = CH_3
X = S; $R_{1,2}$ = Ph

X

Musmanni and Ferraris [121] have prepared the polymer (**XI**), and describe its electrochemical properties.

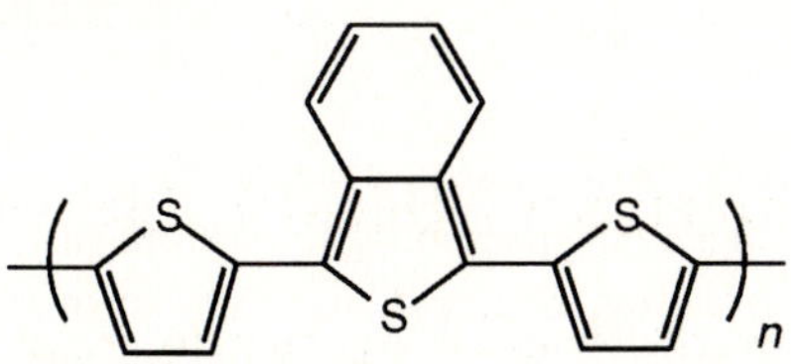

XI

Related to (**XI**) are the compounds based on 5,7-di(2-thienyl)thieno[3,4-b]pyrazine (**XII**) which, when polymerised, yield polymers with low band gaps, that is, λ_{max} is relatively high. The neutral (undoped) polymers are all dark blue/black [122].

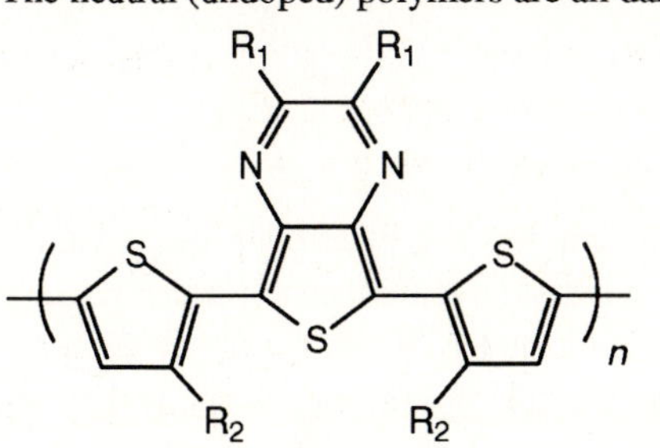

R_1 = R_2 = H
R_1 = Me; R_2 = H
R_1 = C_6H_{13}; R_2 = H
R_1 = $C_{13}H_{27}$; R_2 = H
R_1 = R_2 = $C_{16}H_{13}$

XII

9.4.5 Polymers Derived from Fused-ring Thiophenes

Mastragostino and co-workers [123] prepared the poly(thieno[3,2-b]thiophene) (**XIII**) and studied its electrochromic properties. The ion inserted into the polymer was ClO_4^-, the electrolyte being propylene carbonate containing $LiClO_4$.

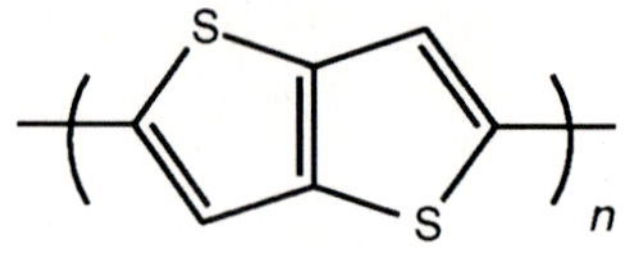

XIII

These workers [123] have also investigated the polymer formed from monomeric dithieno(3,2–b; 2,3–d)thiophene units (**XIV**) used as an electrochrome, and Taliani *et al.* [124] have studied the same polymer.

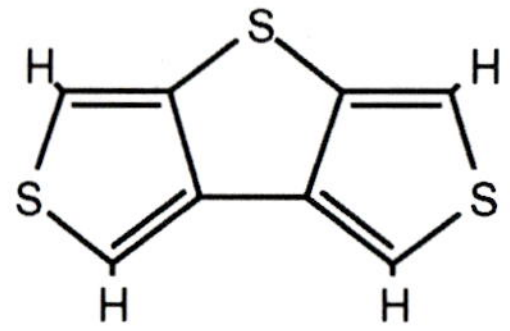

XIV

In contrast to polythiophene, polymeric (**XIV**) is semi-transparent when oxidised (doped) and opaque when undoped. The wavelength maximum for the opaque form is 490 nm [94]; and the response time τ is less than a second, to effect an unspecified optical absorbance change [115]. A cycle life of 1.1×10^4 has been reported [115].

Related to (**XIV**) are the polymers derived by anodic electrochemistry of (**XV**) and (**XVI**), both reported by Gottesfeld *et al.* [112].

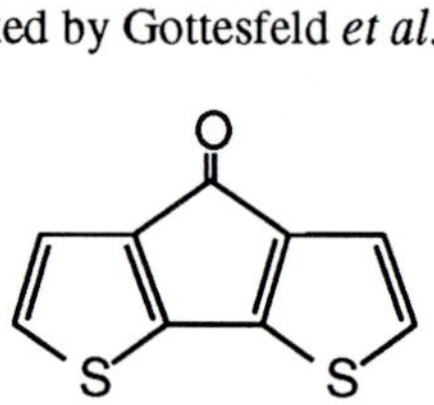

XV

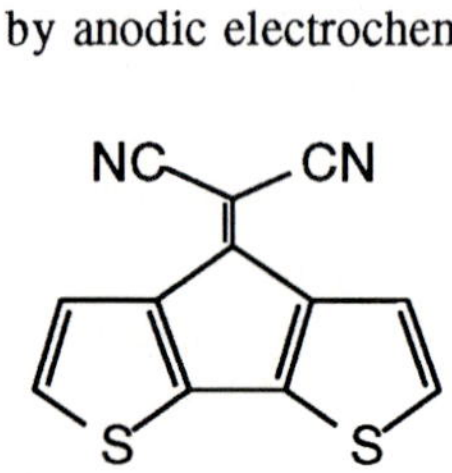

XVI

Heeger and Wudl [125–128] studied the electrochromic switching of poly(isothianaphthene) (**XVII**), formally poly(benzo[c]thiophene), which has a switching time of only a few milliseconds or so [126], and is stable for about 6000 cycles [126]. The polymer is, unusually, yellow when oxidised and black when neutral. Onada *et al.* [129] show the spectra of different redox states of the polymer.

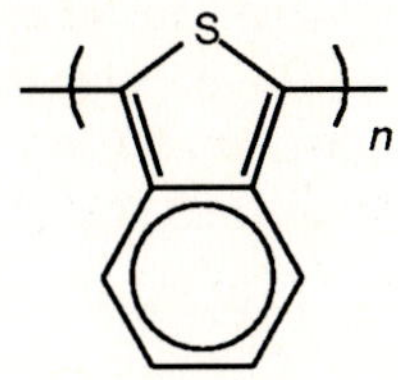

XVII

King and Higgins [130] have prepared a range of related poly(isothianaphthene)s by anodic polymerisation of (**XVIII**). The potential at which reductive doping commenced has a strong dependence on substituent [130]. By contrast, the onset potential for oxidative doping is relatively insensitive to substituent; the compounds made were the 5,6-dichloro, 5-fluoro, 5-methyl and 4-fluoro analogue of (**XVIII**).

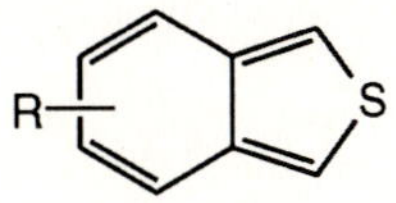

XVIII

Ikenoue has also reported on the electrochromism of poly(naphtho[2,3-c]thiophene) [131]. The change in polymer colour is apparently slight, resulting in a poor *CR*. On oxidative doping, a new optical band is formed at 830 nm and the band at 575 nm becomes less intense. The electrode potential is at 0.69 V *vs* AgCl/Ag.

Kathirgamanathan and Shepherd [132] have shown that poly(phenanthro[9,10c] thiophene) (**XIX**) is electrochromic. This polymer is yellow in the potential range −0.8 to 0 V, and red in the range 0 to 0.8 V.

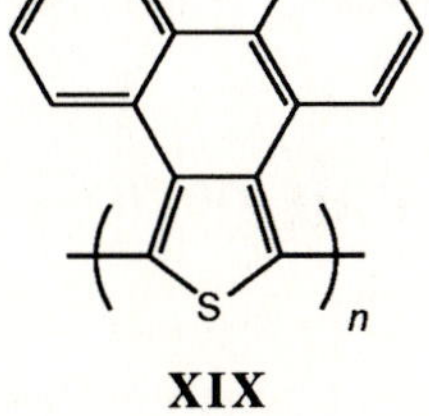

XIX

9.4.6 Polythiophene Copolymers and Composite Materials

The properties of polythiophenes have been further extended by preparing composites of poly(dithienothiophene) or poly(dithienopyrrole) mixed with poly(vinyl chloride) [93].

Oligothiophene co-polymers are short chains of thiophene-containing polymer in which the thiophene units are chemically bonded rather than by physically mixed.

Co-polymers comprising up to eight thiophene units in length, have been prepared with a polymer backbone of either polythene [133] or poly(vinyl chloride) [134].

9.5 Poly(carbazole)

Carbazole (**XX**) can be thought of as a fused-ring pyrrole, although pyrrole polymerises via oxidative coupling at carbon-2 by hydrogen abstraction – a mechanism not available to carbazole species.

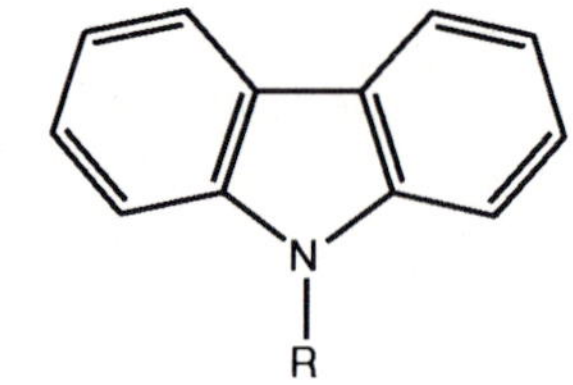

XX

Since straightforward electropolymerisation is not possible, carbazole-based electrochromes are generally prepared by anodic electropolymerisation of *N*-vinylcarbazole monomer [135–137]. Strictly, this type of polymer is not a conducting polymer since monomeric carbazole moieties hang pendant from a polyethene-like backbone.

The doped (reduced) form of polymeric *N*-vinylcarbazole has absorbance maxima at 390 nm and 740 nm. The polymer is stable and undergoes many cycles without degradation [136–137]. The effect of chain length on electrochromism has been reported for polymers of *N*-butyl-3,6-carbazolediyl [138] and *N*-alkyl-3,6-carbazolediyl [139].

A co-polymer of poly(siloxane) and carbazole has been reported by Maud and co-workers [140–143] and a self-doped polymer poly(3,6-(carbaz-9-yl)propane sulphonate) has been prepared by Qui and Reynolds [144].

9.6 Miscellaneous Polymeric Electrochromes

Electrochromic properties have been reported for poly(mercaptohydroquinone) and poly(mercapto-*p*-benzoquinone) [145], poly(*N*-methyl-9,10-dimethyl-phenazasilane [146, 147] poly(phenylquinoxaline) [148] and for poly(2-naphthol) [149].

9.7 Recent Developments

Many conducting polymers have been used in ECDs. Table 9.13 lists those ECDs comprising both organic and inorganic electrochromes. The inorganic electrochrome is generally the primary layer.

Table 9.13 Examples of ECDs Utilising Mixed Organic–Inorganic Electrochromes.

Conducting Polymer Electrochrome	Inorganic Electrochrome	Ref.
polyaniline	Prussian blue	[150]
polyaniline	tungsten trioxide	[151–154]
poly(*N*-benzylaniline)	tungsten trioxide	[155, 156]
poly(methylthiophene)	nickel oxide	[157]
poly(3-methylthiophene)	ITO	[158]
poly(3-octylthiophene)	iridium oxide	[100]
polypyrrole	nickel oxide	[159]

References

[1] A.F. Diaz, K.K. Kanazawa and G.P. Gardini, *J.C.S., Chem. Commun.*, (1979), 635.

[2] G. Wegner, *Mol. Cryst. Liq. Cryst.*, 106 (1984) 269.

[3] M. Gazard, in T.A. Skotheim (ed.), 'Handbook of Conducting Polymers, vol. 1', Marcel Dekker, New York, 1986. Chap. 19.

[4] M. Mastragostino, in B.Scrosati (ed.), 'Applications of Electroactive Polymers', Chapman and Hall, London, 1993. Chap. 7.

[5] K. Hyodo, *Electrochim. Acta*, 39 (1994) 265.

[6] S. Greenberg, in 'Kirk-Othmer Encyclopaedia of Chemical Technology', 4th edn., vol. 6, Wiley, New York, 1993. p. 312.

[7] J.M. Margolis (ed.), 'Conducting Polymers and Plastics', Chapman and Hall, New York, 1989.

[8] R.J. Mortimer, in R.G. Compton and G. Hancock (eds.) 'Research in Chemical Kinetics', vol. 2, Elsevier, Amsterdam, 1994, p 261.

[9] T. Kobayashi, H. Yoneyama and T. Tamura, *J. Electroanal. Chem.*, 161 (1984) 419.

[10] F. Rourke and J.A. Crayston, *J.C.S., Faraday Trans.*, 89 (1993) 295.

[11] B.C. Sherman, W.B. Euler and R.R. Force, *J. Chem. Ed.*, 71 (1994) A94.

[12] T. Hjertberg, W.R. Salaneck, I. Lundstrom, N.L.D. Somasiri and A.G. MacDiarmid, *J. Polym. Sci., Polym. Lett.*, 23 (1985) 503.

[13] T. Kobayashi, H. Yoneyama and H. Tamura, *J. Electroanal. Chem.*, 177 (1984) 281.

[14] V.D. Parker, in M.M. Baizer (ed.), 'Organic Electrochemistry', Marcel Dekker, 1976. Chap. 14.

[15] Y. Wei, G.-W. Jang, C.-C. Chan, K.F. Hsueh, R. Hariharan, S.A. Patel and C.K. Whitecar, *J. Phys. Chem.*, 94 (1990) 7716.

[16] A. Watanabe, K. Mori, Y. Iwasaki, Y. Nakamura and S. Niizuma, *Macromolecules,* 20 (1987) 1793.

[17] S. Kunimura, *Jpn. Kokai Tokkyo Koho., JP*, 62,296,525, (1987), cited in *Chem. Abstr.* 107: P199,892*y*.

[18] Z. Jiang, X. Zhang and Y. Xiang, *J. Electroanal. Chem.*, 351 (1993) 321.

[19] C. Barbero and R. Kötz, *J. Electrochem. Soc.*, 141 (1994) 859.

[20] S. Cordoba-Torresi, A. Hugot-Le Goff and S. Joiret, *Proc. S.P.I.E.*, 1272 (1990) 162.

[21] A. Hugot-Le Goff and S. Cordoba-Torresi, *Proc. Electrochem. Soc.*, 90–2 (1990) 157.

[22] M.-C. Bernard, S. Cordoba-Torresi and A. Hugot-Le Goff, *Solar Energy Mater. Solar Cells*, 25 (1992) 225.

[23] A. Hugot-Le Goff and M.-C. Bernard, *Synth. Met.*, 60 (1993) 115.

[24] A. Efremova, A. Regis and L. Arsov, *Electrochim. Acta,* 39 (1994) 839.

[25] M.-C. Bernard and A. Hugot-Le Goff, *Proc. Electrochem. Soc.*, 94–4 (1994) 185.

[26] M.A. Rodriguez, M.A. De Paoli and M. Mastragostino, *Electrochim. Acta*, 36 (1991) 2143.

[27] M.F. Rubner, in G.J. Ashwell (ed.), 'Molecular Electronics', Research Studies Press, Taunton, 1992. Chap. 2, p. 92.

[28] J.C. Lacroix, K.K. Kanazawa and A. Diaz, *J. Electrochem. Soc.*, 136 (1989) 1308.

[29] A.G. Green and A.E. Woodhead, *J. Chem. Soc., Trans.*, 97 (1910) 2388.

[30] Y. Wei, W.W. Focke, G.E. Wnek, A. Ray and A. G. MacDiarmid, *J. Phys. Chem.,* 93 (1989) 495.

[31] E. Kim, M.-H. Lee, B.S. Moon, C. Lee and S.B. Rhee, *J. Electrochem. Soc.,* 141 (1994) L26.

[32] N. Kobayashi, K. Yamada and R. Hirohashi, *Chem. Lett.*, (1990) 1983.

[33] J. Yano, S. Kai and K. Ogura, *J. Mater. Sci. Lett.*, 12 (1993) 1791.

[34] M.T. Nguyen, R. Paynter and L.H. Dao, *Polymer*, 33 (1992) 214.

[35] T. Ohsaka, M. Ohba, M. Sato, O.N. Makiko, S. Tanaka and S. Nakamura, *Synth. Met.*, 43 (1991) 3089.

[36] T. Ohsaka, M. Ohba, M. Sato, N. Oyama, S. Tanaka and S. Nakamura, *J. Electroanal. Chem.*, 300 (1991) 51.

[37] M.T. Nguyen and L.H. Dao, *J.C.S., Chem., Commun.*, (1990) 1221.

[38] M.T. Nguyen and L.H. Dao, *J. Electroanal. Chem.*, 289 (1990) 37.

[39] J. Guay, R. Paynter and L.H. Dao, *Macromolecules*, 23 (1990) 3598.

[40] V.K. Gator, M.D. Liu, M.D. Love and C.R. Leidner, *J. Electroanal. Chem.*, 257 (1988) 133.

[41] A. Yasuda and J. Seto, *J. Electroanal. Chem.*, 288 (1990) 65.

[42] A.H. Arevalo, H. Fernandez, J.J. Silber and L. Sereno, *Electrochim. Acta*, 35 (1990) 731.

[43] K. Shannon and J.E. Fernandez, *J.C.S., Chem. Commun.*, (1994), 643.

[44] M.A. De Paoli, E.R. Duek and M.A. Rodriguez, *Synth. Met.*, 41 (1991) 973.

[45] S. Kuwabata, M. Susumu, K. Mitsui and H. Yoneyama, *J. Electroanal. Chem.*, 281 (1990) 97.

[46] E.L. Tassi, M.A. De Paoli, S. Panero and B. Scrosati, *Polymer*, 35 (1994) 565.

[47] M. Morita, *Makromol. Chem.*, 194 (1993) 1513.

[48] M. Morita, *Makromol. Chem.*, 194 (1993) 2361.

[49] M. Morita, S. Miyazaki, H. Tanoue, M. Ishikawa and Y. Matsuda, *J. Electrochem. Soc.*, 141 (1994) 1409.

[50] M. Morita and I. Hashida, *J. Polym. Sci.*, B29 (1991) 1567.

[51] M. Morita, *J. Polym. Sci.*, B32 (1994) 231.

[52] S. Kuwabata, N. Takahashi, S. Hirao and H. Yoneyama, *Chem. Mater.*, 5 (1993) 437.

[53] H. Yoneyama and Y. Shoji, *J. Electrochem. Soc.*, 137 (1990) 3826.

[54] R. MacNeill, D.E. Weiss and D. Willist, *Aust. J. Chem.*, 18 (1965) 477.

[55] N.J. Morse, D.R. Rosseinsky, R.J. Mortimer and D.J. Walton, *J. Electroanal. Chem.*, 255 (1988) 119.

[56] J. Lukkari, R. Tuomala, S. Ristimäki and J. Tankara, *Synth. Met.*, 47 (1992) 217.

[57] P.A. Christensen and A. Hamnett, 'Techniques and Mechanisms in Electrochemistry', Blackie, Glasgow, 1994. p. 328.

[58] G.B. Street, T.C. Clarke, R.H. Geiss, V.Y. Lee, A.I. Nazzal, P. Pfluger and J.C. Scott, *J. Phys.*, 44–C3 (1983) 599.

[59] K.K. Kanazawa, A.F. Diaz, R.H. Geiss, W.D. Gill, J.F. Kwak, J.A. Logan, J.F. Rabolt and G.B. Street, *J.C.S., Chem. Commun.*, (1979) 854.

[60] A.F. Diaz and J.I. Castillo, *J.C.S., Chem. Commun.*, (1980) 397.

[61] S. Asavapiriyanont, G.K. Chandler, G.A. Gunawardena and D. Pletcher, *J. Electroanal. Chem.*, 177 (1984) 229.

[62] M. Ogasawara, K. Funahashi, T. Demura, T. Hagiwara and K. Iwata, *Synth. Met.*, 14 (1986) 61.

[63] C.A. Ferreira, S. Aeiyach, M. Delamar and P.C. Lacaze, *J. Electroanal. Chem.*, 284 (1990) 351.

[64] A.F. Diaz and J. Bargon, in T.A. Stockheim (ed.), 'Handbook of Conducting Polymers, vol 1', Marcel Dekker, New York, 1986.

[65] A. Boyle, E. Genies and M. Fouletier, *J. Electroanal. Chem.*, 279 (1990) 179.

[66] D.J. Walton, C.E. Hall and A. Chyla, *Synth. Met.*, 45 (1991) 363.

[67] J.E. Dubois, F. Garnier, G. Tourillon and M. Gazard, *J. Electroanal. Chem.*, 148 (1983) 299.

[68] E.M. Genies, G. Bidan and A.F. Diaz, *J. Electroanal. Chem.*, 149 (1983) 101.

[69] R.C.D. Peres, V.F. Juliano, M.A. De Paoli, S. Panero and B. Scrosati, *Electrochim. Acta*, 38 (1938) 869.

[70] M.A. De Paoli, S. Panero, S. Passerini and B. Scrosati, *Adv. Mater.*, 2 (1990) 480.

[71] M.A. De Paoli, S. Panero, P. Prosperi and B. Scrosati, *Electrochim. Acta*, 35 (1990) 1145.

[72] D.-S. Park, Y.-B. Shim and S.-M. Park, *J. Electrochem. Soc.*, 140 (1993) 2749.

[73] T. Shimizu, T. Kaneko, T. Inoue and H. Takahara, *P.C.T. Int. Appl. WO,* 8,700,843, cited in *Chem. Abstr.* 107: P59,728z.

[74] R.B. Bjorkund and I. Lundström, *J. Electron. Mater.*, 14 (1985) 39.

[75] S.I. Yaniger and D.W. Vidrine, *Appl. Spectrosc.*, 40 (1986) 174.

[76] M. Okano, A. Todo and K. Mochida, *Bull. Chem. Soc. Jpn.*, 63 (1990) 1716.

[77] M. Gazard, *S.P.I.E. International Technical Conference, Geneva, Switzerland,* 1983 cited in ref. [3].

[78] M. Gazard, J.C. Dubois, M. Champagne, F. Garnier and G. Tourillon, *J. Phys. Paris. Colloq.*, C3 (1983) 537, cited in ref. [3].

[79] A.F. Diaz, M. Salmon, J. Addy in, 'Proceedings of the First European Display Research Conference', VDE-Verlag, Berlin, 1981. p. 111.

[80] E.B. Maarouf, D. Billaud and E. Hannecart, *Mater. Res. Bull.*, 29 (1994) 637.

[81] K. Jackowska, A. Kudelski and J. Bukowska, *Electrochim. Acta*, 39 (1994) 1365.

[82] N.M. Hanly, D. Bloor, A.P. Monkman, R. Bonnett and J.M. Ribo, *Synth. Met.*, 60 (1993) 195.

[83] Z. Gao, J. Bobacka, A. Lewenstam amd A. Ivaska, *Electrochim. Acta*, 39 (1994) 755.

[84] J.R. Reynolds, M. Pyo and Y.-J. Qiu, *J. Electrochem. Soc.*, 141 (1994) 35.

[85] S. Ye and D. Bélanger, *J. Electrochem. Soc.*, 141 (1994) L49.

[86] A.E. Hourch, A. Rakotondrainibe, B. Beden, P. Crouigneau, J.-M. Léger, C. Lamy, A.A. Tanaka and E.R. Gonzalez, *Electrochim. Acta*, 39 (1994) 889.

[87] Y. Li and S. Dong, *J. Electroanal. Chem.*, 348 (1993) 181.

[88] Z.Q. Gao, J. Bobacka, A. Lewenstam and A. Ivaska, *Synth. Met.*, 62 (1994) 117.

[89] S. Dong and M. Liu, *Electrochim. Acta,* 39 (1994) 947.

[90] H. Sung, H. So and W. Paik, *Electrochim. Acta*, 39 (1994) 645.

[91] Y. Lin and G.G. Wallace, *Electrochim. Acta,* 39 (1994) 1409.

[92] H. Yoneyama and Y. Shoji, *J. Electrochem. Soc.*, 137 (1990) 3826.

[93] G. Baggiato, G. Casalbore-Miceli, V. Fattori, A. Geri, A. Berlin and G. Zotti, *Synth. Met.*, 57 (1993) 3495.

[94] A.F. Diaz, *Chem. Ser.*, 17 (1981) 145.
[95] J. Wang, *Electrochim. Acta*, 39 (1994) 417.
[96] K. Yoshino, R. Sugimoto, J.G. Rose and W.F. Schmidt, *Jpn. J. Appl. Phys.*, 24 (1985) L33.
[97] K. Yoshino, M. Ozaki and R. Sugimoto, *Jpn. J. Appl. Phys.*, 24 (1985) L373.
[98] E.M. Genies, M. Lapowski, C. Santier and E. Vieli, *Synth. Met.*, 18 (1987) 631.
[99] E.W. Tsai, S. Basok, J.P. Ruiz, J.R. Reynolds and K. Rajeshwar, *J. Electrochem. Soc.*, 136 (1989) 3683.
[100] M.-N. Collomn-Dunand-Sauthier, S. Langlois and E. Genies, *J. Appl. Electrochem.*, 24 (1994) 72.
[101] J. Roncali, G. Garreau, A. Yasser, P. Marque, F. Garnier and M. Lemaire, *J. Phys. Chem.*, 91 (1987) 6706.
[102] A. Galal, A.E. Karagözler, R. Shabana, A. Amer, H. Zimmer and H.B. Mark, *Solar Energy Mater. Solar Cells*, 25 (1992) 339.
[103] J. Roncali, P. Marque, R. Garreau, F. Garnier and M. Lemaire, *Macromolecules*, 23 (1990) 1347.
[104] W.J. Albery, F. Li and A.R. Mount, *J. Electroanal. Chem.*, 310 (1991) 239.
[105] M. Mastragostino and L. Soddu, *Electrochim. Acta*, 35 (1990) 463.
[106] J. Roncali, R. Garreau, D. Delabouglise, F. Garnier and M. Lemaire, *Synth. Met.*, 28 (1989) C341.
[107] M. Lemaire, R. Garreau, J. Roncali, D. Delabouglise, H.K. Youssoufi and F. Garnier, *New. J. Chem.*, 13 (1989) 863.
[108] T. Hagiwara, H. Yamaura, K. Sato, M. Hirasaka and K. Iwata, *Synth. Met.*, 32 (1989) 367.
[109] G.G. Cameron, M.D. Ingram and J.L. Harvie, *Faraday Disc. Chem. Soc.*, 88 (1989) 55.
[110] A. Corradini, M. Mastragostino, A.M. Marinangeli and S. Giacobbe, *Synth. Met.*, 28 (1989) C501.
[111] M. Dietrich and J. Heinze, *Synth. Met.*, 41–43 (1991) 503.
[112] J.P. Ferraris, C. McMackin, D. Torres, D. Meeker, A. Rudge and S. Gottesfeld, *Proc. Electrochem. Soc.*, 94–2 (1994) 195.
[113] D. Adebimpe and P. Kathirgamanathan, *J. Electroanal. Chem.*, 348 (1993) 447.
[114] A. Galal, D.D. Cunningham, A.E. Karagözler, E.T. Lewis, A. Nkansah, A. Burkhardt, O.Y. Ataman, H. Zimmer and H.B. Mark, *Proc. Electrochem. Soc.*, 90–2 (1994) 179.
[115] A. Corradini, A.M. Marinangeli, M. Mastragostino and B. Scrosati, *Solid State Ionics*, 28–30 (1988) 1738.
[116] C. Arbizzani, G. Barbarella, A. Bongini, M. Mastragostino and M. Zambianchi, *Synth. Met.*, 52 (1992) 329.
[117] M. Mastragostino, C. Arbizzani, P. Ferloni and A.M. Marinangeli, *Solid State Ionics*, 53–56 (1993) 471.

[118] M. Mastragostino, C. Arbizzani, A. Bongini, G. Barbarella and M. Zambianchi, *Electrochim. Acta*, 38 (1993) 135.

[119] V. Peulon, G. Barbey, O.F. Gerard and C. Paulmier, *Synth. Met.*, 53 (1993) 115.

[120] T. Mitsuhara, S. Tanaka and K. Kaeriyama, *J.C.S., Chem. Commun.*, (1987) 764.

[121] S. Musmanni and J.P. Ferraris, *J.C.S., Chem. Commun.*, (1993) 173.

[122] C. Kitamura, S. Tanaka and Y. Yamashita, *J.C.S., Chem. Commun.*, (1994), 1585.

[123] M. Mastrogostino, A.M. Marinangeli, A. Corradini and C. Arbizzani, *Electrochim. Acta,* 32 (1987) 1589.

[124] C. Taliani, G. Ruani, R. Zamboni, A. Bolognesi, M. Catellani, S. Destri, W. Porzio and P. Ostoja, *Synth. Met.*, 28 (1989) C507.

[125] A.J. Heeger, J. Wudl, J.H. Kaufman and T.-C. Chung, *J. Electrochem. Soc.*, 131 (1984) 2092.

[126] H. Yashima, M. Kobayashi, K.-B. Lee, D. Chung, A.J. Heeger and F. Wudl, *J. Electrochem. Soc.*, 134 (1987) 46.

[127] M. Kobayashi, N. Colaneri, M. Boysel, A.J. Heeger and F. Wudl, *J. Chem. Phys.*, 82 (1985) 5717.

[128] A.J. Heeger, F. Wudl, N. Colaneri and M. Kobayashi, *Synth. Met.*, 14 (1986) 45.

[129] M. Onoda, H. Nakayama, S. Morita and K. Yoshino, *J. Electrochem. Soc.*, 141 (1994) 338.

[130] G. King and S.J. Higgins, *J.C.S., Chem. Commun.*, (1994), 825.

[131] Y. Ikenoue, *Synth. Met.*, 35 (1990) 263.

[132] P. Kathirgamanathan and M.K. Shepherd, *J. Electroanal. Chem.*, 354 (1993) 305.

[133] K. Nawa, K. Miyawaki, I. Imae, N. Noma and Y. Shirota, *J. Mater. Chem.*, 3 (1993) 113.

[134] M. Wakabayashi, M. Masaki, K. Yuichi, O. Takaki, K. Osakada and T. Yamamoto, *Synth. Met.*, 57 (1993) 3632.

[135] J.E. Dubois, P.C. Lacaze and A. Desbène-Monvernay, *J. Electroanal. Chem.*, 132 (1982) 177.

[136] J.E. Dubois, P.C. Lacaze, A. Desbène-Monvernay, P.L. Desbène, J.J Basselier and D. Richard, *J. Electroanal. Chem.*, 147 (1983) 107.

[137] J.E. Dubois, P.C. Lacaze and A. Desbène-Monvernay, *J. Electroanal. Chem.*, 152 (1983) 87.

[138] Y. Pelous, G. Froyer, D. Ades, C. Chevrot and A. Siove, *Polym. Commun.*, 31 (1990) 341.

[139] A. Siove, D. Ades, E. N'Gbilo and C. Chevrot, *Synth. Met.*, 38 (1990) 331.

[140] D.M. Goldie, A.R. Hepburn, J.M. Maud and J.M. Marshall, *Synth. Met.*, 55 (1993) 1650.

[141] A.R. Hepburn, J.M. Marshall and J.M. Maud, *Synth. Met.*, 43 (1991) 2935.

[142] D.M. Goldie, A.R. Hepburn, J.M. Maud and J.M. Marshall, *Mol. Cryst., Liq. Cryst.*, 236 (1993) 87.

[143] J.M. Maud, A. Vlahov, D.M. Goldie, A.R. Hepburn and J.M. Marshall, *Synth. Met.*, 55 (1993) 890.

[144] Y.J. Qui and J.R. Reynolds, *J. Electrochem. Soc.*, 137 (1990) 900.

[145] G. Arai, K. Imai, K. Tanaka and I. Yasu, *Bull. Chem. Soc. Jpn.*, 64 (1991) 505.

[146] G. Casalbore-Miceli, A. Geri, G. Giro, S. Daolio and G. Zotti, *Synth. Met.*, 40 (1991) 317.

[147] G. Casalbore-Miceli, G. Baggiato, S.S. Emmi, A. Geri, S. Daolio, L. Favaretto and D. Pietropaolo, *J. Appl. Electrochem.*, 20 (1990) 989.

[148] F. Templier, J. Torres, A. Halimaouli, J. Palleau and J.C. Oberlin, *Proc. Electrochem. Soc.*, 91–1 (1991) 712.

[149] M.C. Pham, P.C. Lacaze, F. Genoud, L.H. Dao and M. Nguyen, *J. Electrochem. Soc.*, 140 (1993) 912.

[150] E.A.R. Duek, M.A. De Paoli and M. Mastragostino, *Adv. Mater.*, 5 (1993) 650.

[151] B.P. Jelle and G. Hagen, *Proc. Electrochem. Soc.*, 94–2 (1993) 324.

[152] B.P. Jelle and G. Hagen, *J. Electrochem. Soc.*, 140 (1993) 3560.

[153] B.P. Jelle and G. Hagen, *Electrochim. Acta*, 38 (1993) 1497.

[154] B.P. Jelle, G. Hagen, S. Sunde and R. Oedegaard, *Synth. Met.*, 54 (1993) 315.

[155] M.T. Nguyen and L.H. Dao, *J. Electrochem. Soc.*, 136 (1989) 2131.

[156] M.T. Nguyen and L.H. Dao, *Proc. Electrochem. Soc.*, 90–2 (1990) 246.

[157] C. Arbizzani, M. Mastragostino, S. Passerini, R. Pileggi and B. Scrosati, *Electrochim. Acta*, 36 (1991) 837.

[158] A. Corradini, A.M. Marinangeli and M. Mastragostino, *Electrochim. Acta*, 35 (1990) 1757.

[159] C. Avino, S. Panero and B. Scrosati, *J. Mater. Chem.*, 3 (1993) 1259.

10 Other Organic Electrochromes

10.1 Monomeric Species

10.1.1 Carbazoles

Related to the poly(*N*-vinylcarbazole) in chapter 9 are films of radical-cation salt formed by one-electron oxidation of monomeric substituted carbazole species (**I**). In their neutral form, carbazoles are soluble and colourless, whereas films of radical cation generated oxidatively form a highly coloured precipitate on the electrode. Results obtained for several carbazoles by Dubois and co-workers [1] are summarised in table 10.1.

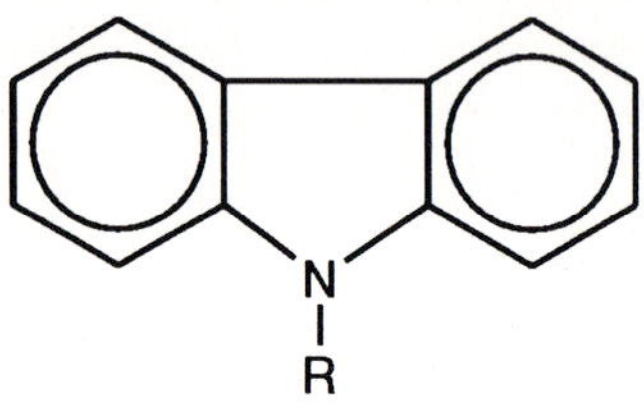

I

Table 10.1 Colours and Electrode Potentials of Polymers derived from various Carbazoles in MeCN solution (ref. [1]).

Carbazole Monomer	Colour of Radical Cation	$E^{o}_{(1)}$/V
Carbazole	dark green	+0.9
N-ethylcarbazole	green	+1.3
N-phenylcarbazole	'iridescent'	+1.2
N-carbazylcarbazole	yellow brown	+1.1

10.1.2 Methoxybiphenyl Compounds

A different class of compounds are the violenes [2] based on a core of polymethoxybiphenyl. The uncharged parent compounds are colourless, but form a film of brilliantly coloured radical-cation salt following electro-oxidation.

Many of the compounds were prepared by Parker and co-workers, who used an interesting anodic coupling reaction [3]. They also studied the compounds electrochemically [4] although the most comprehensive study of their electrochromism is that by Grant *et al.* [5].

Many of these species should more correctly be called biphenyls, fluorenes or phenanthrenes according to the nature of the bridging group (if any) between the two rings. The stability of the radical cation formed by one-electron oxidation of the neutral species is a function of molecular planarity, as is shown by the stability series (**II**) ≪ (**III**) < (**IV**), where compound (**II**) is forced out of planarity by the steric repulsion induced by the two *o*-methoxy groups, whereas (**IV**) is always planar owing to the methylene bridge.

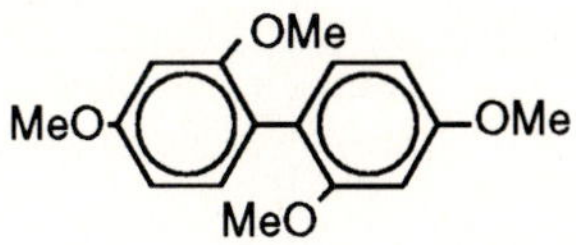

II

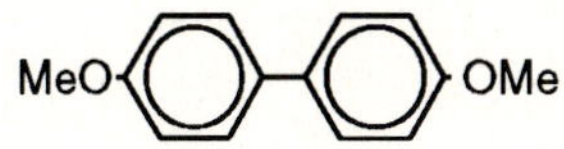

III

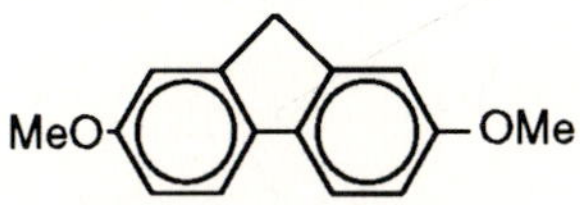

IV

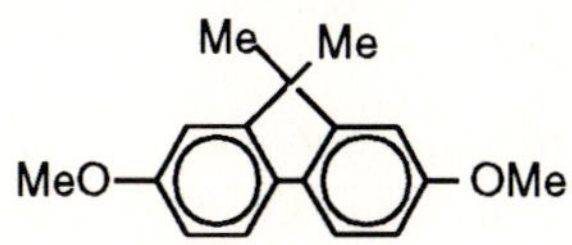

V

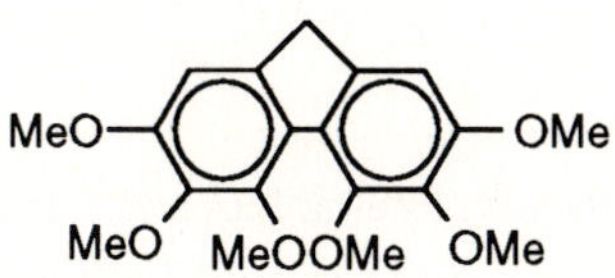

VI

As a crude generalisation [5], fluorenes with one methoxy group show irreversible oxidations, but the electrochemistry of compounds with two or more methoxy groups is found to be more reversible, with *ortho* and *meta* methoxy substituents exhibiting lower redox potentials [5]. The fluorene compounds which appear to be most suitable for ECD inclusion, that is, those yielding the stablest films of radical-cation salt, are listed in table 10.2.

Other fluorene compounds investigated did not form radical cations of sufficient stability for viable use as electrochromes, or evinced irreversible electrochemistry. Some biphenyl compounds of interest are listed in table 10.3, within which compound (**VII**) deprotonates slowly to form 2,7-dimethoxyphenanthrene.

Table 10.2 Colours, CV Peak Potentials and Spectral Properties for Methoxybiphenyl Species Forming Solid Radical-Cation Films on Reduction in MeCN Solutions (ref. [5]).

Compound	Colour of Radical	$E_{pa(1)}$/V	$E_{pc(1)}$/V	λ_{max}/nm	ε/dm^3 mol^{-1} cm^{-1}
(IV)	Blue	+0.91	+0.79	411	40,400
(V)	—	+0.96	+0.84	385	32,800
(VI)	Blue	+0.87	+0.81	415	44,300

(**IV**) is 2,7-dimethoxyfluorene. (**V**) is 2,7-dimethoxy-9,9-dimethylfluorene. (**VI**) is 2,3,4,5,6,7-hexamethylfluorene.

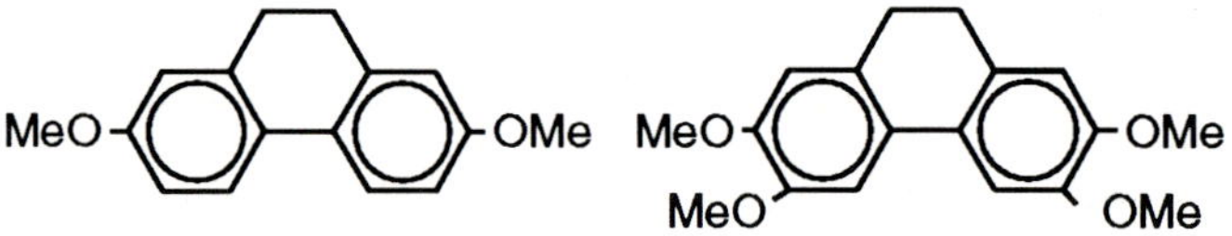

VII **VIII**

Table 10.3 Colours, CV Peak Potentials and Spectral Properties for Methoxybiphenyl Species Forming Only Soluble Radical-Cation on Reduction in dichloromethane-TFA (5:1) solution (ref. [4]).

Compound	Colour of Radical	$E_{pa(1)}$/V	$E_{pc(1)}$/V	λ_{max}/nm	ε/dm^3 mol^{-1} cm^{-1}
(III)	—	+1.28	+1.22	417	29,512
(VII)	Green	+1.14	+1.07	386	20,420
(VIII)	Green	+0.94	+0.88		

(**III**) is 2,7-dimethoxybiphenyl. (**VII**) is 2,7-dimethoxy-9,10-dihydrophenanthrene. (**VIII**) is 2,3,6,7-tetramethoxy-9,10-dihydrophenanthrene.

10.1.3 Quinones

Many quinones are soluble, stable and only moderately coloured as neutral molecules but, on one-electron reduction, form brightly-coloured stable solid films of radical anion on the electrode [6–10].

Benzoquinones (both *ortho* (**IX**) and *para* (**X**)), 1,4-naphthaquinones (**XI**), and anthra-9,10-quinones (**XII**) have been studied, for example, compound (**X**) in propylene carbonate containing $LiClO_4$ [8]. Table 10.4 lists electrode potentials and spectroscopic data for electrochromic quinone-based systems.

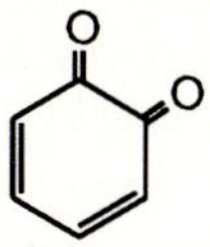

IX

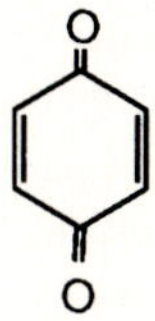

X

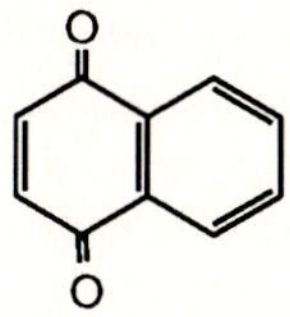

XI

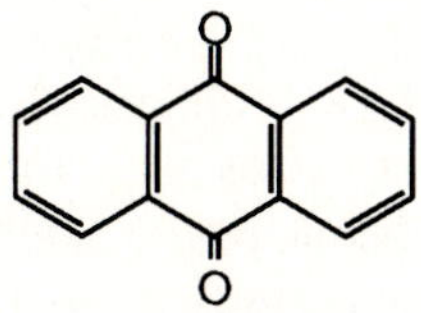

XII

Table 10.4 Quinone Systems: Film-forming Properties, Colours, Wavelength Maxima, and Reduction Potentials (E_{pc} from CVs, or standard electrode potentials E^{o}; all solutions in MeCN with 0.1 M TEAP).

Quinone (RQ)	Solid Film?	Colour of $RQ^{-\bullet}$	λ_{max}/ nm	$E_{pc(1)}$/ V	$E_{pc(2)}$/ V	Ref.
o-3,4,5,6-tetrachlorobenzoQ	yes	intense blue		−0.170	+0.210	[6, 7]
o-3,4,5,6-tetrabromobenzoQ	yes	blue		−0.190	+0.140	[6]
p-benzoQ	yes	light blue		−0.720	−0.430	[6, 8]
p-2,3,5,6-tetrafluorobenzoQ	no	yellow		−0.430	−0.100	[6]
p-2,3,5,6-tetrachlorobenzoQ	no	yellow		−0.420	−0.060	[6]
p-2,3-dicyano-5,6-dichloro-benzoQ	no	pink		+0.070	+0.330	[6]
5-aminonaphthoQ	yes	purple/blue	410	$E^{o}_{(1)} = -0.83$		[9]
1-aminoanthraQ	yes	—	—	$E^{o}_{(1)} = -1.03$		[9]
2-aminoanthraQ	yes	—	—	$E^{o}_{(1)} = -0.99$		[9]
1,5-diaminoanthraQ	yes	purple	570	$E^{o}_{(1)} = -1.10$		[9]

In general, *ortho* benzoquinones are more electrochemically stable than their *para* isomers, *o*-chloranil (3,4,5,6-tetrachlorobenzoquinone) being the most stable quinone studied, having a cycle life in excess of 10^5 cycles [6].

Aminoanthraquinones show a more complicated electrochemical behaviour: at moderate potentials, two redox couples are exhibited during cyclic voltammetry, representing first quinone^0 $\rightleftharpoons$ quinone$^{-\bullet}$, then quinone$^{-\bullet}$ $\rightleftharpoons$ quinone^{2-} at more negative potentials. In addition to this behaviour, polymerisation of the amine occurs when the electrode is made very positive, in a similar manner to the formation of polyaniline (chap. 9). Only monomeric species have been listed in table 10.4.

A trichromic ECD [11] with the capacity to form the colours red, green and green/blue, has been developed using 2-ethylanthraquinone in propylene carbonate together with 4,4'-*bis*(dimethylamino)-diphenylamine. The anthraquinone compound produces the red colouration when reduced (CR = 2:1 at λ_{max} = 545 nm) and diphenylamine provides the other colours in two different oxidation states: its first oxidation product is a green radical cation (CR = 2:1) and a subsequent oxidation product is a green/blue dication (CR = 3.5:1 at $\lambda_{max} \approx$ 500 nm).

10.1.4 Diphenylamine and Phenylene Diamines

Würster salts [12–14], the mono-oxidised *p*-phenylene diamines, are highly coloured, and are thus candidate electrochromes.

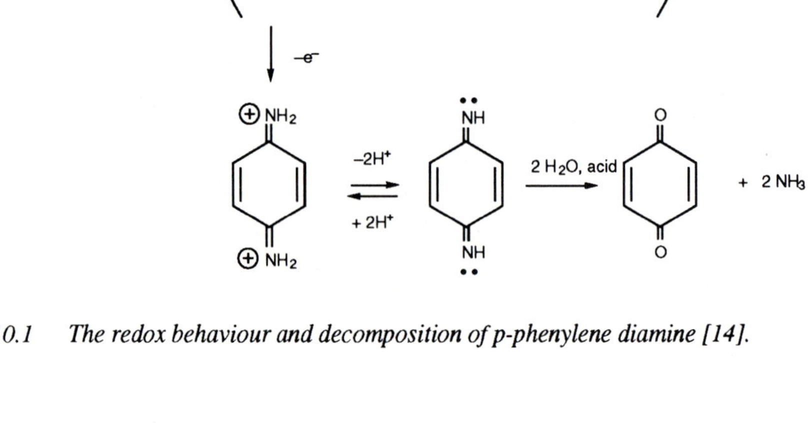

Fig 10.1 The redox behaviour and decomposition of p-phenylene diamine [14].

The oxidation proceeds as in Fig. 10.1, two oxidation peaks appearing in CVs in aqueous Na_2SO_4. The first product is stable because of the resonance shown, but the second decomposes to quinone with H^+ present. Alkylated members as the *N,N,N',N'*-tetramethyl-*p*-phenylene diamine give stabler radical cations, which are blue or blue/green.

10.1.5 Miscellaneous Monomeric Electrochromes

A trichromic ECD has been fabricated including 2,4,5,7-tetranitro-9-fluorenone as the red-forming material, 2,4,7-trinitro-9-fluorenylidene malononitrile as the green and TCNQ as the blue electrochrome [15]. The electrochemistry of the 2,4,7-trinitro-9-fluorenylidene malononitrile species is reported in reference [16].

10.2 Tethered Electrochromic Species

10.2.1 Pyrazolines

A tethered organic system which has received some attention is that based on the oxidation of the pyrazolines (**XIII**) and (**XIV**), spectral details for which are listed in table 10.5.

Kaufman *et al.* [17, 18] have published most of the current work on tethered pyrazolines. Such species are more intensely absorbing than the TTF species below, and have faster response times τ [18]. Pure pyrazoline monomers are readily prepared, and are soluble in many solvents prior to polymerisation [18].

A solid state ECD which incorporates such polymeric pyrazolines has recently been constructed [19], and a response time of 10 ms and a *CR* of 10 are cited.

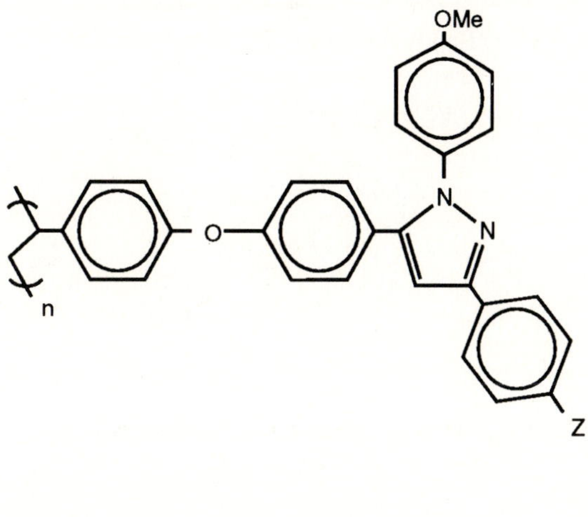

XIII Z is —OMe

XIV Z is

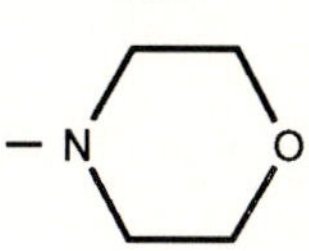

Table 10.5 Half-Wave Potentials, Colours and Response Times τ for Tethered Pyrazoline Species in MeCN containing 0.1 M TEAP electrolyte (ref. [17]).

Compound	$E_{1/2}$/V	Colour change	λ_{max}/nm	τ/ms
(XIII)	+0.55	yellow-to-green	510	50
(XIV)	+0.45	yellow-to-red	554	100

10.2.2 Tetracyanoquinodimethane (TCNQ)

Neutrally-charged TCNQ is a stable, colourless molecule that forms a coloured radical anion following one-electron reduction. The stability of this radical is ascribed to appreciable delocalisation of the single negative charge via the ring over the CN groups.

Since TCNQ and its radical anion are both soluble in most common solvents, Chambers [20–22] improved the electrochromic write–erase efficiency by chemically affixing the TCNQ species (**XV**) to an electrode surface by means of polymerisation.

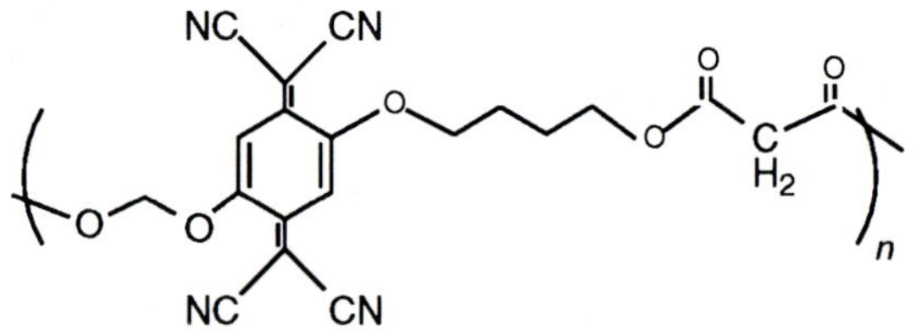

XV

Table 10.6 Spectroscopic Data for TCNQ Redox Species in MeCN solution (ref. [21]).

Species	λ_{max}/nm	ln(ε/dm^3 mol^{-1} cm^{-1})
$TCNQ^{o}$	408	5.06
	430	5.06
$TCNQ^{-\bullet}$	445	4.30
	660	3.38
	728	3.92
	812	4.20

The oligomer (**XV**) is estimated [20] to have a molecular weight of about 2200. Electrodes modified with (**XV**) are electrochemically reversible [20]. Spectroscopic data for TCNQ and $TCNQ^{-\bullet}$ are listed in table 10.6.

In solution, additional species to those in table 10.6 have also been identified, including a di-anion $(TCNQ)^{2-}$ and (in aqueous solution only) a dimer di-anion $(TCNQ)_2^{2-}$ [21].

10.2.3 Tetrathiafulvalene (TTF)

Like TCNQ, TTF has been used in ECDs chemically tethered to an electrode surface. Thus, Kaufman [23–25] used the two species (**XVI**) and (**XVII**) to modify electrodes. In early trials, a TTF device underwent > 10^4 cycles without deterioration [23]. The electrochromic TTF colouration accompanies oxidation of neutral TTF to form a radical cation; spectral characteristics of (**XVI**) and (**XVII**) are listed in table 10.7.

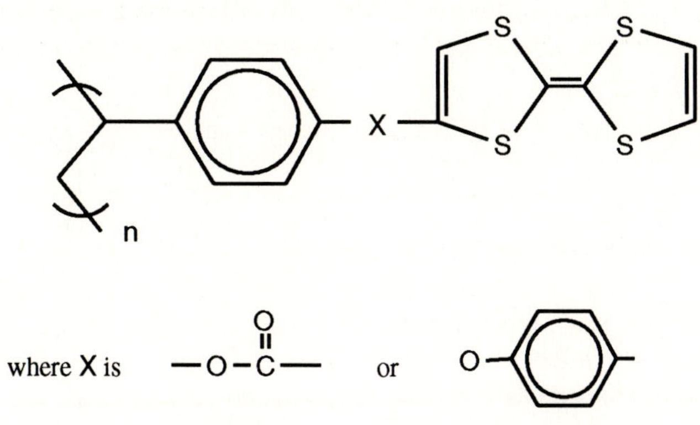

XVI **XVII**

Table 10.7 Half-wave Potentials, Colours, Wavelength Maxima and Response Times τ for Tethered TTF Species, (ref. [18]).

Compound	$E_{1/2}$/V	Colour change	λ_{max}/nm	τ/ms[a]
(**XVI**)	+0.45	orange-to-brown	515	200
(**XVII**)	+0.35	yellow-to-green	650	150

[a] Time required for a charge injection of 1 mC cm^{-2} into a film of thickness 5 μm.

The response of these TTF-based devices is relatively slow. Electrochemical studies [18, 24] show the rate-determining step to be ion movement into the film during colouration [26]; furthermore, electron transport through the film proceeds via hopping or tunnelling between TTF sites.

In addition to $TTF^{+\bullet}$, the other TTF species listed in table 10.8 will also form in the layer around the electrode; their spectral characteristics are reproduced here in table 10.8. Although the minor species in table 10.8 do not contribute much to the colouration of a TTF device, they greatly complicate the electrochemistry.

Table 10.8 Spectroscopic Data for TTF Redox Species in MeCN Solution ([23]).

Species	λ_{max}/nm
$TTF^{+\bullet}$	393, 653
$(TTF^{+\bullet})_2$	1800
$(TTF)_2^{+\bullet}$	820
TTF^{2+}	533

Recent TTF displays comprise solid-state devices with polymeric electrolytes [19].

10.3 Electrochromes Immobilised by Viscous Solvents

Dissolution or dispersion of an electrochrome in an electrolyte of high viscosity allows enhancement of the write–erase efficiency. Such immobilised species are type 3 electrochromes. The usual matrix for entrapment is an electrolyte gel of high viscosity [28] such as polyelectrolytes or polymeric electrolytes.

Methylene blue (**XVIII**) is electrochromic being blue when oxidised; the leucomethylene blue formed on reduction is colourless. (**XVIII**) and all its other redox states are soluble in polar solvents but may be immobilised by dissolution in semi-solid polyAMPS [29] or in polyaniline [30]. Clearly, only a small proportion of the electrochrome dispersed will be electroactive.

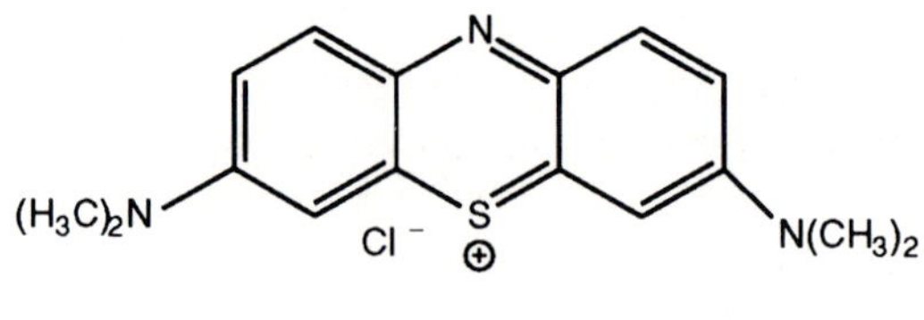

XVIII

Carbazoles (*cf.* sections 9.5 and 10.1 above) have been immobilised similarly, using polysiloxane as the viscous electrolyte [31–34].

References

[1] J.E. Dubois, F. Garnier, G. Tourillon and M. Gazard, *J. Electroanal. Chem.*, 129 (1981) 229.

[2] S. Hünig, *Pure Appl. Chem.*, 15 (1967) 109.

[3] A. Ronlán, O. Hammerich and V.D. Parker, *J. Am. Chem. Soc.*, 95 (1973) 7132.

[4] A. Ronlán, J. Coleman, O. Hammerich and V.D. Parker, *J. Am. Chem. Soc.*, 96 (1974) 845.

[5] B. Grant, N.J. Clecak, M. Oxsen, A. Jaffe and G.S. Keller, *J. Org. Chem.*, 45 (1980) 702.

[6] A. Desbène-Monvernay, P.C. Lacaze and A. Cherigui, *J. Electroanal. Chem.*, 260 (1989) 75.

[7] J.E. Dubois, A. Desbène-Monvernay, A. Cherigui and P.C. Lacaze, *J. Electroanal. Chem.*, 169 (1984) 157.

[8] M. Yashiro and K. Sato, *Jpn. J. Appl. Phys.*, 20 (1981) 1319.

[9] V.K. Gater, M.D. Liu, M.D. Love and C.R. Leidner, *J. Electroanal. Chem.*, 257 (1988) 133.

[10] V.K. Gater, M.D. Love, M.D. Liu and C.R. Leidner, *J. Electroanal. Chem.*, 235 (1987) 381.

[11] T. Ueno, Y. Hirai and C. Tani, *Jpn. J. Appl. Phys.*, 24 (1985) L178.

[12] A. Watanabe, K. Mori, Y. Iwasaki, Y. Nakamura and S. Niizuma, *Macromolecules*, 20 (1987) 1793.

[13] L. Michaelis, M.P. Schubert and S. Gramick, *J. Am. Chem. Soc.*, 61 (1939) 1981.

[14] L. Michaelis, *Chem. Rev.*, 16 (1935) 243.

[15] A. Yasuda and J. Seto, *J. Electrochem. Soc.*, 136 (1989) 419C, abstract number 634.

[16] A. Yasuda and J. Seto, *J. Electroanal. Chem.*, 303 (1991) 161.

[17] F.B. Kaufman and E.M. Engler, *J. Am. Chem. Soc.*, 101 (1979) 547.

[18] F.B. Kaufman, A.H. Schroeder, E.M. Engler and V.V. Patel, *Appl. Phys. Lett.*, 36 (1980) 422.

[19] Y. Hirai and C. Tani, *Appl. Phys. Lett.*, 43 (1983) 704.

[20] R.W. Day, G. Inzelt, J.F. Kinstle and J.Q. Chambers, *J. Am. Chem. Soc.*, 104 (1982) 6804.

[21] G. Inzelt, R.W. Day, J.F. Kinstle and J.Q. Chambers, *J. Electroanal. Chem.*, 161 (1984) 147.

[22] G. Inzelt, R.W. Day, J.F. Kinstle and J.Q. Chambers, *J. Phys. Chem.,* 87 (1983) 4592.

[23] F.B. Kaufman, *Conference Record of the I.E.E.E., Biennial Display Research Conference, New York,* 1978, p. 23.

[24] F.B. Kaufman and E.M. Engler, *J. Am. Chem. Soc.,* 101 (1979) 739.

[25] F.B. Kaufman, J.B. Torrance, J.A. Scott, B. Welber and P. Seiden, *Phys. Rev.,* B19 (1979) 730.

[26] F.B. Kaufman, A.H. Schroeder, E.M. Engler, S.R. Kramer and J.Q. Chambers, *J. Am. Chem. Soc.,* 102 (1980) 483.

[27] N. Ohnishi, K. Kondo and K. Takemoto, *Makromol. Chem.*, 191 (1990) 2397, cited in *Chem. Abstr.* 114: 8471 *x.*

[28] H. Tsutumi, Y. Nakagawa, K. Miyazaki, M. Morita and Y. Matsuda, *J. Polym. Chem.*, 30 (1992) 1725.

[29] J.M. Calvert, T.J. Manuccia and R.J. Nowak, *J. Electrochem. Soc.*, 133 (1986) 951.

[30] S. Kuwabata, K. Mitsui and H. Yoneyama, *J. Electroanal. Chem.*, 281 (1990) 97.

[31] A.R. Hepburn, J.M. Marshall and J.M. Maud, *Synth. Met.*, 43 (1991) 2935.

[32] D.M. Goldie, A.R. Hepburn, J.M. Maud and J.M. Marshall, *Mol. Cryst. Liq. Cryst.*, 236 (1993) 87.

[33] J.M. Maud, A. Vlahov, D.M. Goldie, A.R. Hepburn and J.M. Marshall, *Synth. Met.*, 55 (1993) 890.

[34] D.M. Goldie, A.R. Hepburn, J.M. Maud and J.M. Marshall, *Synth. Met.*, 55 (1993) 1650.

Part III

Elaborations

11 Polyelectrochromism†

11.1 Introduction

Uses may be envisaged in which one or more electrochromes evince a whole series of different colours, in states generated at the corresponding values of applied potential. For a single-species electrochrome, a series of oxidation states, or charge states, each coloured, could be produced each at a particular potential (if each such state can be sustained, that is, if the species is 'multivalent' in one dialect of chemical parlance). The vanadium-ion system in say dilute aqueous sulphuric acid is a solution-phase example [1]: the colour sequence and approximate potentials [1] with respect to SCE are lavender V^{2+} (–0.5 V), sea-green V^{3+} (+0.1 V), blue VO^{2+} (0.5 V) and bronze-yellow $VO_2{}^+$ (+0.9 V). The oxycations here result from reaction with H_2O, by abstraction of O^{2-} with the liberation of H^+. These colours, ionic forms and potentials are of course specific to vanadium ions in aqueous sulphate solution; such multivalence of metal ions is not uncommon but not universal (a minority are single-valent, like Na^+ or Ca^{2+}). A comparable system best envisaged as organic in nature could involve the bonding together of several of the same (two-state) electrochrome as sub-units, which, if there were some electron delocalisation (linkage) in a general sense between the sub-units, could show sequential colouration during a potential scan. (In the absence of such interaction, all sub-units would show the same response to potential, all undergoing colouration simultaneously).

Mixed electrochromic systems with each component having its own switching potential could similarly effect multiple colour/potential responses. However, if a particular single-species colour is to be evoked, then it has to be arranged that the other species are in their colourless states, otherwise the several colours will superimpose, in a more or less foreseeable manner as long as the discrete states arising with the variation of potential do not chemically interact.

So far, the experiments envisaged here involve single-electrode systems, where the total electrochrome ensemble is subject to the same applied potential. Another configuration of possible utility comprises individually addressed (connected) micro-electrodes (pixels) in arrays to effect different regions or dots of colour by applying differentiated potentials to the individual pixels. Such a configuration requires insulation from each other of not only the metal-contact points, but also, in the case of all-solid electrochromes, of the attached deposits of electrochrome. Dissolved electrochromes could not be used in such multiple electrode systems, for the following reason, which also exemplifies the need for encapsulation of the islets of *solid* deposit if capable of mutual interaction where in contact. If say V^{2+} were generated at one point electrode, and VO^{2+} at an adjacent one then (in seconds or minutes in this example, perhaps more slowly in others) the reaction [2]

† In our view *Electropolychromism* is better.

$$V^{2+} + VO^{2+} (+ 2H^+) \rightarrow 2 V^{3+} (+ H_2O) \quad (11.1)$$

would ensue. With other choices of potential for neighbouring micro-electrodes, the reactions [2]

$$V^{2+} + VO_2^+ (+ 2H^+) \rightarrow V^{3+} + VO^{2+} (+ H_2O) \quad (11.2)$$

or

$$V^{3+} + VO_2^+ \rightarrow 2 VO^{2+} \quad (11.3)$$

could occur, where the first and last of these three reactions are *comproportionations* [2] and the intervening one is a variant of this class. Comproportionation is always a possibility in systems comprising a single electrochrome material in three (or more) oxidation states (if E^o values permit). This complication is a special nuisance for systems impregnated into paper, as is envisaged below, in sections 11.2.1, 11.2.3 and 12.3.1.

11.2 Studies of Polyelectrochromic Systems

11.2.1 Bipyridiliums

The polyelectrochromism accessible in bipyridilium systems arises from the reactions

$$\underset{\text{(usually colourless)}}{bipm^{2+}} + e^- \rightarrow \underset{\text{(colour 1)}}{bipm^{+\bullet}} \quad (11.4)$$

$$\underset{\text{(colour 1)}}{bipm^{+\bullet}} + e^- \rightarrow \underset{\text{(colour 2)}}{bipm^{o}} \quad (11.5)$$

where, depending on the substituents on nitrogen, colour 2 can be made either strong or weak. Detail is presented in chapter 8. The electrochromic processes of methyl viologen and of *N*-phenylhydroxy-*N'*-methyl viologen have been studied with the electrochromic material loaded into paper, as matrix [5].

11.2.2 Polybipyridyl Systems

The three colours evinced by bipyridilium units appear to be extendable to multiply-coloured species, by linking several units together. As has been remarked, for polyelectrochromism, electron uptake by one unit needs to be communicated to the others, by intervening delocalisation within the connecting electronic network forming the bonding system. Such delocalisation would alter the energy levels of hitherto unaffected units so that a different applied potential would be required for subsequent electron uptake by one of the latter; the differing energy levels would then effect the evocation of a new colour on this subsequent electron uptake. The argument applies doubly to linked bipyridilium units, since each can undergo uptake of a *second* electron. Studies have been pursued with this in mind, as follows.

The molecules (**I**) and (**II**) have been synthesised by straightforward methods [3].

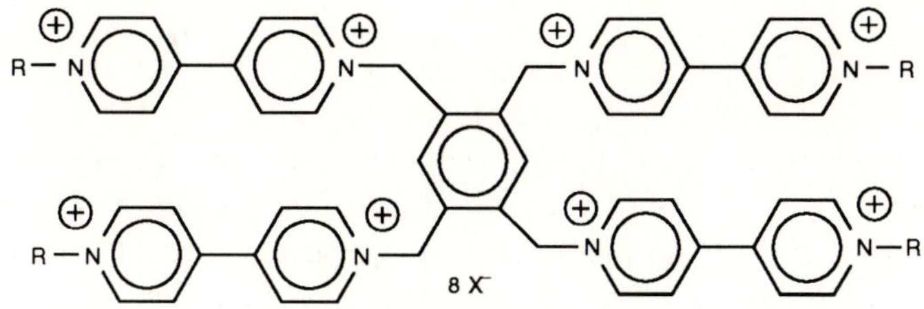

I (R = Me) and **II** (R = Bz)

The maximum expected number of colours for *n* linked bipm units is $2n + 1$, here 9. The CV of (**I**), in the outcome, showed only three peaks (Fig 11.1), corresponding to only three new colours, of species (**III**) to (**V**).

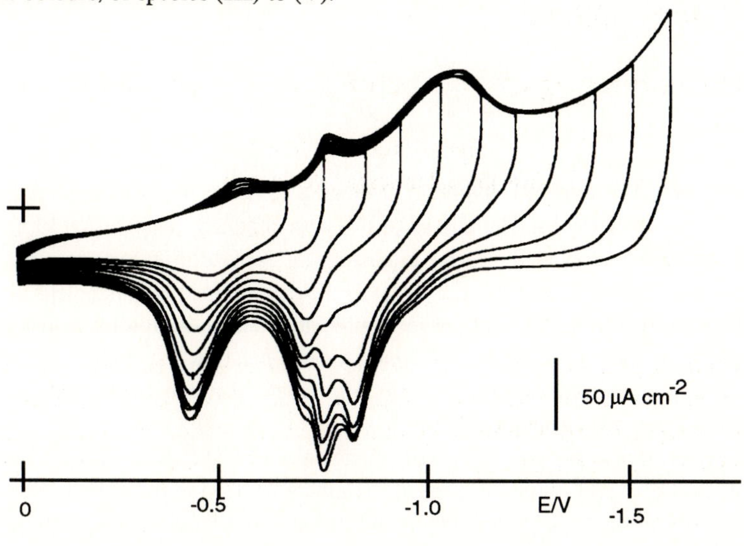

*Fig 11.1 The CVs (proceeding clockwise) of (**I**) $4Br^- 4I^-$ (10^{-3} M) in aq. KCl (0.1 M) as a function of scan limit E_λ. The scan rate v was 50 mV s^{-1} throughout. (Figure reproduced from ref. [3] with permission of Chapman and Hall.)*

Considerations of relative currents implied that the following three states correspond to the observed colours and peaks (with assumed pairing of radical-cation electrons indicated by double arrows):

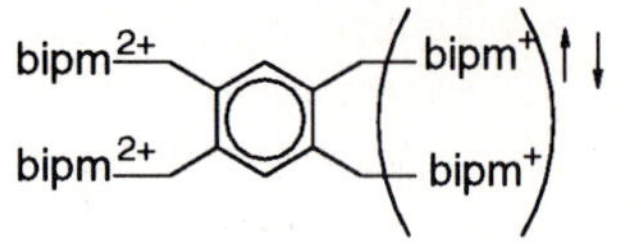

III

(red)

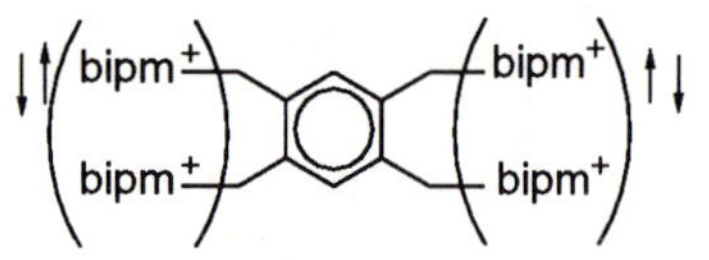

IV

(purple)

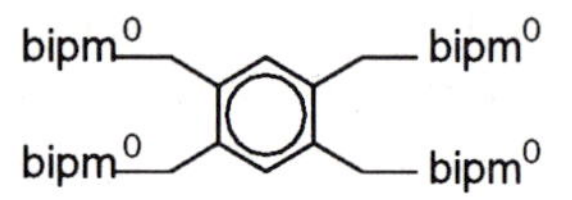

V

(fawn)

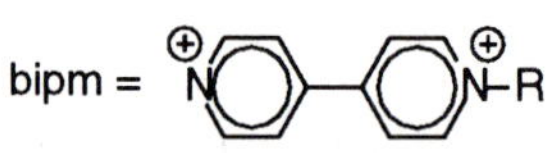

where (**II**) → (**III**) involved uptake of 2 electrons, (**III**) → (**IV**) two electrons, and (**IV**) → (**V**), four electrons. The colour of (**V**) was subdued, as for usual bipm^{O} states, but (**III**) and (**IV**) showed the usual intensities of bipyridilium systems. It might be concluded that inter-unit interaction is virtually absent, but the particular steps of electron acquisition imply a different conclusion: appreciable delocalisation leads to a molecule undergoing gross steps of 2-electron + 2-electron + 4-electron acquisitions in succession, each resulting in a particularly stable total-molecule electron configuration possibly arising from the indicated electron pairings.

These particular molecules offer little advantage in ECD applications but do indicate the need of weaker (but not zero) inter-unit interactions, for multicoloured responses. An EC system colouring on oxidation would undergo electron loss(es) in a parallel sequence of steps; no candidate for a polyelectrochromic system made up of linked oxidisable units appears to have been studied.

11.2.3 Metal Hexacyanometallates

The core compound (dating back to Diesbach, 1704) is Prussian blue (PB) or ferric hexacyanoferrate(II), which can be partly oxidised to Berlin green (BG), wholly oxidised to

'Prussian brown' (PX) – actually yellow in thin-film form – or reduced to Prussian white (PW, a clear compound). Thus [4]

$$\underset{\text{(PB)}}{[\mathrm{Fe^{III}\,Fe^{II}\,(CN)_6}]^-} \rightarrow \underset{\text{(BG)}}{[\mathrm{Fe^{III}\,Fe^{II+\delta}\,(CN)_6}]^{(1-\delta)-}} + \delta e^- \qquad (11.6)$$

$$\underset{\text{(PB)}}{[\mathrm{Fe^{III}\,Fe^{II}\,(CN)_6}]^-} \rightarrow \underset{\text{(PX)}}{[\mathrm{Fe^{III}\,Fe^{III}\,(CN)_6}]^0} + e^- \qquad (11.7)$$

$$\underset{\text{(PB)}}{[\mathrm{Fe^{III}\,Fe^{II}\,(CN)_6}]^-} + e^- \rightarrow \underset{\text{(PW)}}{[\mathrm{Fe^{II}Fe^{II}\,(CN)_6}]^{2-}} \qquad (11.8)$$

(Various other trivial names have been given).

Other metal hexacyanometallates have been studied: together with the PB systems, these have been detailed in chapter 6. All the electrochromic processes of PB have been further studied with the electrochrome impregnated into paper together with barely moist electrolyte [5].

11.2.4 Phthalocyanines

As an example of these commonly polyelectrochromic systems, the lutetium electrochrome $Lu(pc)_2$ (chap. 5) is initially brilliant green, on oxidation going yellow/tan, and yet more anodically, red. On reduction of the green compound, blue first ensues then a violet/blue product [6, 7]. The probable structures are given in chapter 5, together with further examples.

11.2.5 *Tris*(dicarboxyester-2,2'-bipyridine) Ruthenium Systems

A monomeric metal complex, the *tris*(3-acrylatoprop-1-oxyl)-2,2'-bipyridine) ruthenium(II) *p*-toluenesulphonate, spin coated onto an OTE and thermally polymerised, gives an impressive seven-colour electrochromic system covering the spectral range, which can be scanned in 250 ms. Good stability is claimed [8].

11.2.6 Mixed Systems

When both anode and cathode are electrochromic, dual colouration ensues (e.g. ref. [9]). A novel mixture of a bipyridilium and a metal cyanometallate system has been shown to extend the range of colours which can be evolved electrochromically at *one* electrode [9].

Thus PB on ITO glass has been covered by a Nafion® (sulphonated polytetrafluoroethane polymer) film as electrolyte, on to which a layer of methyl viologen has been deposited. As the potentials (*vs* SCE) for the PX-to-PW switches range from +0.9 V to 0.0 V, while MV^{2+}, $MV^{+\bullet}$ occurs at −0.8 V, the purple colour of $MV^{+\bullet}$ (actually $MV^{+\bullet}$, blue, with some dimer, red) is now added to the range of the PB system, as shown in Fig. 11.2. Comparable bilayer systems comprise PB–tungsten trioxide [10] and PB–polyaniline [11]. The scope for elaboration is open-ended since following [9], PB could in principle be substituted by other cyanometallates, while many $bipm^{2+}$ systems could replace MV^{2+} to operate over the negative end of the potential range.

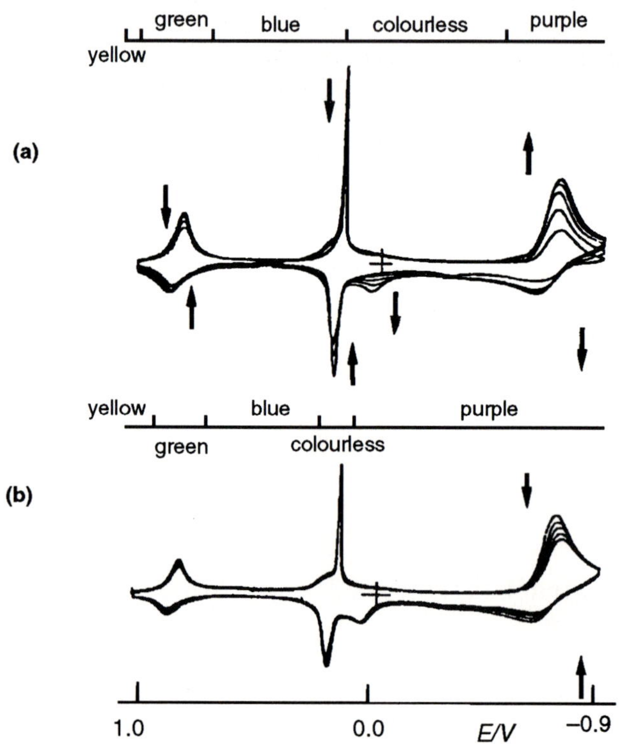

Fig. 11.2 *CVs recorded at 10 mV s^{-1} directly after immersion of OTE (with 1.2 × 10^{-3} C cm^{-2} Prussian blue film and 2.14 × 10^{-7} mol cm^{-2} Nafion® outer layer) in solution: (a) 10^{-4} M methyl viologen and 0.2 M KCl and (b) 0.2 M KCl after transfer from solution (a). Potential scans started at +0.5 V and reversed at −0.9 V and +1.0 V. The arrows indicate the way peak currents vary with increasing scan number. The colours evoked at any particular potential are indicated by the lines above each CV. (After ref. [9].)*

References

[1] L.G. Sillén and A.E. Martell, 'Stability Constants of Metal-Ion Complexes', The Chemical Society, London, 1964. Table 1; W.D. Bare and W. Resto, *J. Chem. Ed.*, 71 (1994) 692.

[2] H. Taube, J.E. Sutton and P.M. Sutton, *Inorg. Chem.*, 18 (1979) 1017.

[3] D.R. Rosseinsky and P.M.S. Monk, *J. Appl. Electrochem.*, in press.

[4] R.J. Mortimer and D.R. Rosseinsky, *J.C.S., Dalton Trans.*, (1984) 2059.

[5] D.R. Rosseinsky and J.L. Monk, *J. Electroanal. Chem.*, 270 (1989) 473.

[6] A.T. Chang and J.C. Marchon, *Inorg. Chim. Acta,* 53 (1981) L241.

[7] M.M. Nicholson, *Ind. Eng. Chem., Prod. Res. Develop.*, 21 (1982) 261.

[8] C.M. Elliott and J.G. Redepenning, *J. Electroanal. Chem.*, 197 (1986) 219.

[9] R.J. Mortimer, *J. Electrochem. Soc.*, 138 (1991) 633.

[10] K. Honda, M. Fujita, H. Ishida, R. Yamamoto and K. Ohgaki, *J. Electrochem. Soc.*, 135 (1988) 3151.

[11] E.A.R. Duek, M.-A. De Paoli and M. Mastragostino, *Adv. Mater.*, 5 (1993) 650.

12 Photoelectrochromism and Electrochromic Printing

12.1 Introduction and Definitions

Systems which change colour electrochemically, but only on being illuminated, are termed *photoelectrochromic* (*cf. electrochromic* or *photochromic* [1] when only one of these stimuli is applied). Only a few photoelectrochromic systems have been examined as such, although in some studies of photoelectrochemistry, colour changes are mentioned [2–4].

12.1.1 Mode of Operation

Two bases of photoelectrochromic operation are available. In the first, the potential required to evoke electrochromism is already applied but can act only through a photo-activated switch, filter or trigger. A separate photoconductor or other photocell could serve as switch, or the actual electrochromic electrode surface itself could be a photoconductor, or sandwiched together with a photoconductor. Such *photo-activated* systems contrast with *photo-driven* devices, in which illumination of one or other part of the circuit produces the photovoltaic potential required to drive the electrochromic current.

12.1.2 Direction of Beam

The direction of illumination is important. During cell operation, if the incident beam traverses a (minimum) distance in the cell prior to striking the photoactive layer, then illumination is said to be 'front-wall' [5], as shown by arrow (a), Fig. 12.1. Conversely 'back-wall' illumination, arrow (b), Fig. 12.1, operates with the beam directed from behind the cell so traversing more cell material before reaching the photo-sensitive layer. Front-wall illumination generally yields superior results since additional absorptions by other layers within the ECD are minimised. Back-wall illumination is used only if undesirable photolytic processes occur with front-wall illumination of the cell.

12.2 Device Types

12.2.1 Devices Containing a Photocell

The simplest circuits for photoelectrochromic device operation comprise a conventional *electrically*-driven ECD together with a photo-operated switch. The switch operated by illumination of a suitable photocell, be it photovoltaic or photoconductive, could trigger a micro-processor or similar element which in turn switches on the already 'poised' cell.

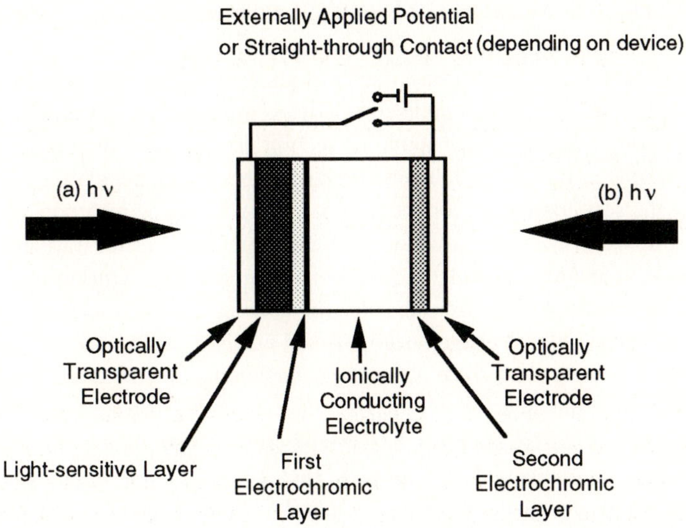

Fig. 12.1 Schematic representation of a photoelectrochromic cell. Illumination from direction (a) represents front-wall illumination: (b) back-wall illumination.

Such an arrangement is not intrinsically photoelectrochromic but is switched on by photocontrolled circuitry: the cell itself could be any of the electrochromic systems in Part II.

12.2.2 Devices Containing Photoconductive Layers

Photoconductive materials are insulators in the absence of light but become conductive when illuminated. Such photoconductors are usually semiconductors like amorphous silicon but, in recent years, many new organic photoconductors have become possible candidates. The mechanism of photoconduction involves the photoexcitation of charge carriers (electrons or holes) from localised sites into the delocalised energy levels forming the conduction band. Such charge so mobilised can be driven by an externally applied potential [6], giving current which can effect electrochromism.

Since photoconduction can occur only at illuminated areas, photoelectrochromic *images*, rather than uniform blocks of tone, may be formed when the device is illuminated through a patterned mask or a photographic negative [7]; electrochromism proceeds only where the conducting parts of the substrate can act on the electrochrome. Polyaniline, which can function as both photoconductor and electrochrome, has been used by itself in a sandwich assembly somewhat like Fig. 12.1, illuminated through a photographic negative. Images so formed on a film had remarkable clarity. Reversal of the cell polarity bleached the image in the absence of light [8, 9]. Silicon has also been used as a photoconductor, with thin-film polypyrrole as the electrochrome [7, 10].

Electrochromic cells may employ a layer of photoconductive material in one of two ways [11, 12]. The first, using a photoconductive component *outside* the ECD, involves a photocell switch as described above in section 12.2.2: illumination of the photoconductor completes the circuit, allowing for electrochromic colouration, which ceases in the dark. In the second method, a photoconductive layer may be incorporated *within* the electrochromic cell. Fig. 12.2 shows an ECD with a photoconductor positioned between an optically conducting substrate and the electrochrome film. During electrochromic colouration or bleaching, ions from the electrolyte enter the electrochromic layer in the usual way (section 2.5), but electrons are injected via the photoconductor. This arrangement has the difficulty that, since most photoconductors are somewhat opaque, ECDs operating with a photoconductor will probably have to operate in a reflective mode. Back-wall illumination of the ECD in Fig. 12.2 would allow for strong metallic electrodes to be employed as the photoconductor support. The photoconductor might conceivably be located between the electrochrome and the electrolyte layers (Fig. 12.3) [11, 13]. Here the photoconductor would need to be ion-permeable: note that the attendant physical stresses of continual ion movement through the photoconductor could lead to eventual conductor disintegration, and so the arrangement in Fig. 12.2 is preferred.

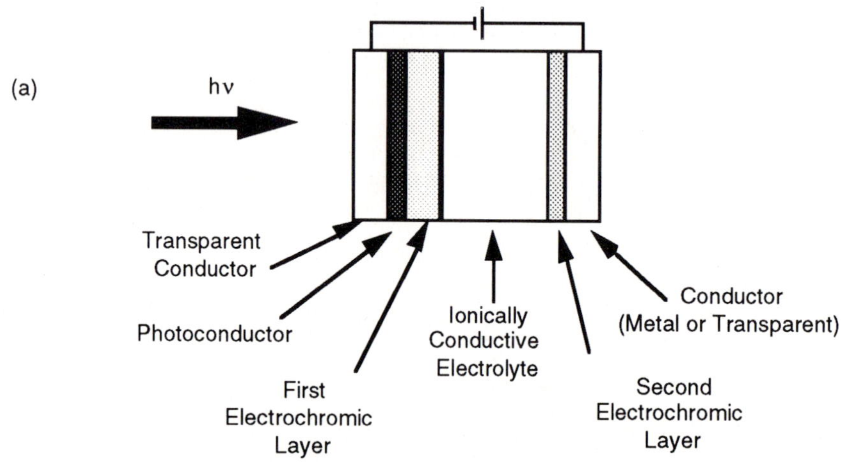

Fig. 12.2 Front-wall illumination of an ECD containing a photoconductive layer between the transparent conductor and the primary electrochrome layer.

12.2.3 Cells Containing Photovoltaic Materials

A photovoltaic material produces a *potential* when illuminated, from a process similar to the excitation of electrons within a photoconductor but with an internal rectifying field which provides a driving force on the electrons. The ionic charge accompanying the electrochromism enters the film from the electrolyte. The photovoltaic layer is not consumed in this process.

The photovoltage required can be quite small since the actual magnitude is not a problem. For example the cell can be 'poised' with an external bias applied, of a voltage that itself is too small to cause the required redox chemistry to occur. Illumination of this poised cell generates a photovoltage which, supplementing the external bias, is sufficient to cause electron transfer now to proceed. For example, WO_3 on TiO_2 is photoelectrochromic, but requires a small bias [14] since the photovoltage generated by illumination is insufficient. Other photoelectrochromic cells operating via photovoltaism include tungsten trioxide on CdS [13, 15], GaAs [16] or GaP [17]. Prussian blue (PB) has also been used in photoelectrochromic devices, with either polycrystalline *n*-type $SrTiO_3$ [18, 19] or CdS [20] as the photolayer. (Indeed, PB has been used with WO_3 to make a photo-rechargeable battery [20].)

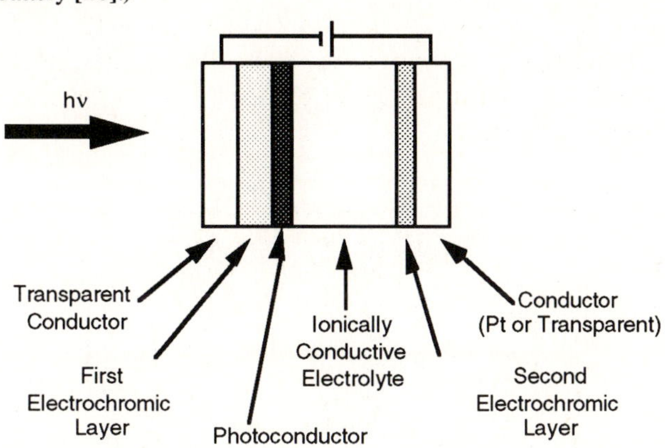

Fig. 12.3 Front-wall illumination of an ECD containing a photoconductive layer between the primary electrochrome layer and the electrolyte.

12.2.4 Cells Containing Photogalvanic Materials

Photogalvanic materials generate current when illuminated. Since net photochemical reactions occur, the photogalvanic material is consumed during the photoreaction [13]: the (photo-operated) write–erase efficiency will therefore be poor. Photoelectrochromism in the cell WO_3 | PEO, H_3PO_4 (MeCN) | V_2O_5 is believed to operate in a photogalvanic sense [13] since the brown colour of the V_2O_5 layer disappears gradually during continual illumination. Curiously, the cell is still photoelectrochromic even after the colour of the V_2O_5 has gone and an alternative cathodic reaction (possibly catalysed oxygen consumption, or VO_2 reduction?) must be envisaged.

12.2.5 Electrochemically Fixed Photochromic Systems

The five-ring compound [21] **Ia** (scheme 12.1) can be photoswitched forward (to the six-ring **Ib** with a 312 nm UV band, and back (to **Ia**) with 600 nm visible light, in a quite standard photochromic on/off system (Fig. 12.5). The novel property exploited here is the electrochemical conversion of **Ib** to photo-inert compound **II**, which 'fixes' the photochromic 'on' state against the photo-switching, unless the reverse electroreduction back to **Ib** is effected. Thus photochromic writing with UV (**Ia** → **Ib**) can be safeguarded by electro-oxidation (**Ib** → **II**) to a state *which can still be read* (in fact **II** absorbs more strongly than **Ib**) and finally after electroreductive unlocking (**II** → **Ib**), the information may be erased by red light.

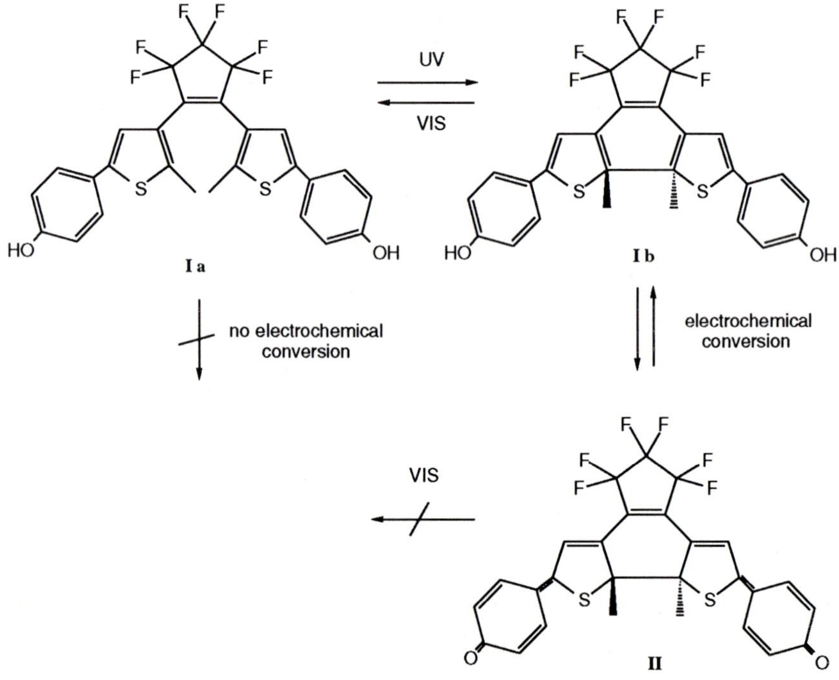

Scheme 12.1

(While clearly not a simple photoelectrochromic system, two separate processes being involved in the operation as outlined, ordinary photoelectrochromism would ensue if **Ia** were illuminated *with the oxidative potential applied,* resulting immediately in the absorbance enhancement missing in **Ib** *cf.* **II**). Cyclic voltammograms are shown in Fig. 12.4.

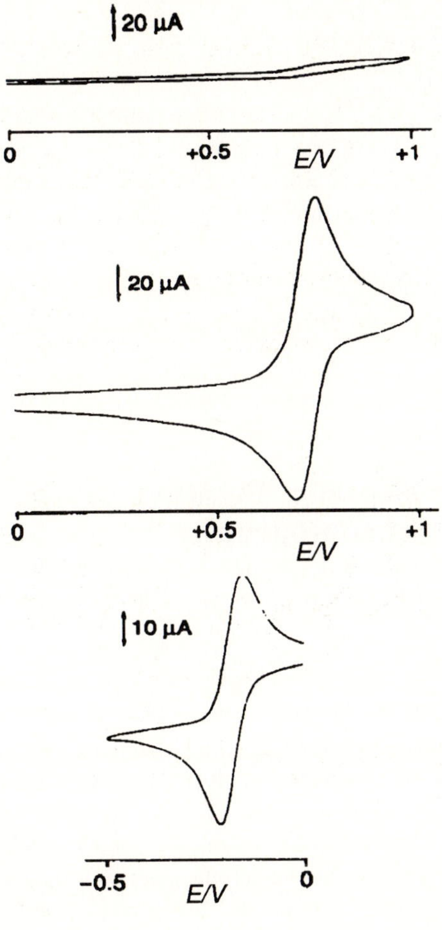

*Fig. 12.4 Cyclic voltammograms determined for compound **Ia** (top), **Ib** (middle) and **II** (bottom), all in MeCN solution containing TBAT electrolyte. (Figure reproduced from ref. [21] with permission of the Royal Society of Chemistry.)*

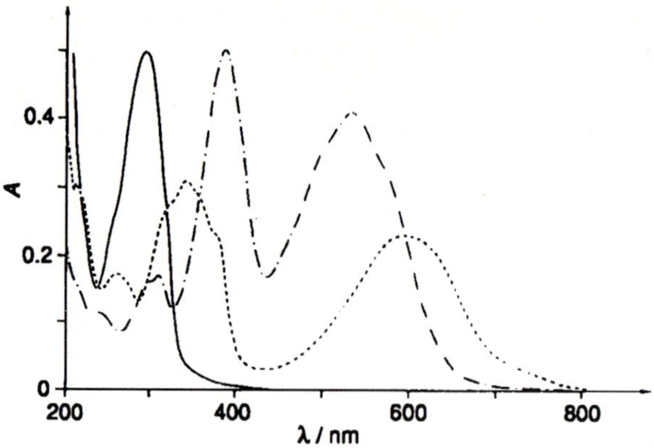

Fig. 12.5 *Absorption spectra of compounds **Ia** (full line), **Ib** (dotted line) and the quinone compound **II** (partially-dashed line) at 10^{-5} M in MeCN. (Figure reproduced from ref. [21] with permission of the Royal Society of Chemistry.)*

12.3 Electrochromic Printing or Electrochromography

12.3.1 Introduction: Monochrome Printing

Photoelectrochromism has been shown above to be effective in imprinting an image on an electroactive polymer, by allowing current flow through a photoconductor to a substrate only where illuminated parts of a projected image allow it. Clearly a digitised image could similarly be printed by appropriately directing voltage pulses to the corresponding pixel electrodes in a multi-pixel array, in contact with paper incorporating an electrochrome and electrolyte, the non-image side of which rests intimately on a conducting surface comprising the counter electrode. Patents going back to 1918 have been issued for such system [22, 23], but the earliest one is Bain's of 1843 [24, 25].

A non-technical account has been given [24] of just such a system, first used in primitive fax-like transmissions in the 19th century. The paper impregnated with ferrocyanide was fed out damp from a roll while the pen in constant contact comprised an iron point, anodic potential pulses on which produced blackened imprints on the paper resting on a metal sheet as counter electrode. Prussian blue might have been expected from this process but it is known [26] that calcined PB (i.e. admixed with iron oxide) gives a black colouration, much as observed. According to the report, the British Meteorological Office still uses just such a system for transmitting cloud-cover diagrams, so belying the disparagement [24] of the substrate as 'soggy electrolytic paper'. In paper of marginal moistness, the electrochemistry of both Prussian blue and viologens ('electrochromography') can be quite impressively reproduced, as though in an electrochemical cell [27].

A large number of recent patents has been issued for elaborations of electrochromic printing systems usually based on organic electrochromic dyes [28]. Electropolymerisation (chap. 9), of monomer such as aniline or pyrrole impregnated into paper, via a stylus electrode or an electrode array, creates black print or images [29].

12.3.2 Polyelectrochromic Printing: Single Electrochromes

Because of comproportionation and similar intervalence interactions outlined in section 12.1, the substrate for polyelectrochromic printing needs to be dot-impregnated with separate islets of electrochrome. While it is possible to imagine effecting polychromatic colouration in a multi-pixel device, resulting from the evocation of corresponding colours by specifically addressed applied potentials, the availability of a single electrochrome to so respond seems problematic – none is known at present. Thus a mixture of electrochromes must be considered.

12.3.3 Four-colour Printing with Mixed Electrochromes

Standard colour printing processes require the four colours cyan, magenta, yellow and black. However, for a direct 'positive' printing system, it is necessary to use a *subtractive* process [30], in which originally coloured materials present in the paper are selectively bleached, and the process of electro-bleaching is involved, as a variant of electrochromism. In effect, four successive colour-filtered images are needed to provide four corresponding different patterns of voltage-imprints onto the pixel electrodes, each *bleaching* one particular colourant to leave the appropriate colour in the surviving combination of colourants. The magenta (reddish hues), cyan (blue) and yellow often need black or dark brown also, to confer definition. To effect this bleaching electrochromically, a specific sequence of voltage pulses is required. Thus a high positive potential could electrobleach one colourant (which must undergo a following spontaneous chemical reaction to stay bleached no matter what further potentials are impressed: this requires an 'ec' mechanism, meaning, here, electrochemical-followed-by-chemical reaction [31]). A high negative potential could electrobleach a second colourant, again with an 'ec' response. What of the third and fourth colourants? These will have been micro-encapsulated in low-melting wax capsules within the paper, by a well established technology, to be now released by quick warming or a dose of microwave radiation. The liberated third colourant may be electrobleached by a low positive potential, with 'ec' response as before, and the fourth colourant by a low negative potential again with 'ec' response.The lowness of the potentials to be used in steps 3 and 4 leaves the remaining colourants 1 and 2 unaffected, as is required. Thus four-colour electrochromic printing is possible, in concept at least.

Direct photocopying would require the use of a photoconductor capable of effecting the four decolourisation steps as just described, at the four differing applied potentials. Clearly some basic research is needed to establish colourants having the required electrochemical responses, and an appropriate photoconductor to withstand the electrochemistry proceeding on its surface [32] will also need to be sought. There seems to be no barrier in principle to achieving these desiderata.

References

[1] H.G. Heller, in L.S. Miller and J.B. Mullin (eds.), 'Electronic Materials: From Silicon to Organics', Plenum Publishing Co., New York, 1991.

[2] K. Hirochi, M. Kitabataka and O. Yamazaki, *J. Electrochem. Soc.*, 133 (1986) 1973.

[3] W. Buttner, P. Rieke and N.R. Armstrong, *J. Electrochem. Soc.*, 131 (1984) 226.

[4] M.P. Stilkans, Y.Y. Purans and Y.K. Klyavin, *Zh. Tekh. Fiz.*, 61 (1991) 91.

[5] R.D. Rauh, *Stud. Phys. Chem.*, 55 (1988) 277.

[6] J.A. Duffy, 'Energy Levels and Bands in Inorganic Solids', Longmans, Harlow, 1990. J. Guillet, 'Polymer Photophysics and Photochemistry', Cambridge University Press, Cambridge, 1987.

[7] O. Inganäs and I. Lundström, *Synth. Met.*, 21 (1987) 13.

[8] H. Yoneyama, N. Takahashi and S. Kuwabata, *J.C.S., Chem. Commun.*, (1992) 716.

[9] H. Yoneyama, *Adv. Mater.*, 5 (1993) 394.

[10] O. Inganäs and I. Lundström, *J. Electrochem. Soc.*, 131 (1984) 1129.

[11] M. Shizukuishi, S. Shimizu and E. Enoue, *Jpn. J. Appl. Phys.*, 20 (1981) 2359.

[12] H. Yoneyama, K. Wakamoto and H. Tamura, *J. Electrochem. Soc.*, 132 (1985) 2414.

[13] P.M.S. Monk, J.A. Duffy and M.D. Ingram, *Electrochim. Acta*, 38 (1993) 2759.

[14] B. Ohtani, T. Atsumi, S. Nishimoto and T. Kagiya, *Chem. Lett.*, (1988) 295.

[15] M. Stilkans, J. Kleparis and E.J. Klevins, *Latv. P.S.R. Zinat. Akad. Vestis. Fiz. Tekh. Zinat. Ser.*, 4 (1988) 43.

[16] B. Reichman, F-R. F. Fan and A.J. Bard, *J. Electrochem. Soc.*, 127 (1980) 333.

[17] M.A. Butler, *J. Electrochem. Soc.*, 131 (1984) 2185.

[18] J.P. Ziegler, E.K. Lesniewski and J.C. Hemminger, *J. Appl. Phys.*, 61 (1987) 3099.

[19] J.P. Ziegler and J.C. Hemminger, *J. Electrochem. Soc.*, 134 (1987) 358.

[20] M. Kaneko, T. Okada, H. Minoura, T. Sugiura and Y. Ueno, *J. Electrochem. Soc.*, 35, (1990) 291.

[21] S.H. Kawai, S.L. Gilat and J.-M. Lehn, *J.C.S., Chem. Commun.*, (1994) 1011.

[22] P.S. Hana, Netherlands Pat. 5,142 (1920).

[23] U. Schmieschek and F. Klutke, German Pat. 684,619 (1939).

[24] T. Hunkin, *New Scientist*, 13th February 1993, p. 33.

[25] A. Bain, UK Pat. 27th May 1843.

[26] Mm. Riffault, Vergnaud and Toussaint, ed. M.F. Malepeyre, translated by A.A. Fesquet, 'A Practical Treatise on the Manufacture of Colours for Printing', Sampson Low, Marston, Low and Searle, London, 1874, p. 531.

[27] D.R. Rosseinsky and J.L. Monk, *J. Electroanal. Chem.*, 270 (1989) 473.

[28] J.E. Kassner, *J. Imaging Technology*, 12 (1986) 325.

[29] R.D. Balanson, G.A. Corker and B.D. Grant, *IBM Technical Disclosure Bulletin*, 26 (1983) 2930.

[30] R.W.G. Hunt, 'The Reproduction of Colour', Fountain Press, Tolworth, England, 1987, pp. 29 and 573.

[31] A.J. Bard and L.R. Faulkner, 'Electrochemical Methods: Fundamentals and Applications', Wiley, New York, 1980, p. 430.

[32] D.R. Rosseinsky and F.R. Mayers, *J.C.S., Dalton Trans.*, (1990) 3419.

Index

(T) refers to a Table in the text